POSTGENOMICS

POSTGENOMICS

Perspectives on Biology
after the Genome

Sarah S. Richardson and Hallam Stevens, editors

DUKE UNIVERSITY PRESS *Durham and London* 2015

Typeset in Whitman
by Westchester Publishing Services

Library of Congress Cataloging-in-Publication Data
Postgenomics : perspectives on biology after the
genome / Sarah S. Richardson and Hallam Stevens,
editors.
pages cm
Includes bibliographical references and index.
ISBN 978-0-8223-5922-7 (hardcover : alk. paper)
ISBN 978-0-8223-5894-7 (pbk. : alk. paper)
ISBN 978-0-8223-7544-9 (e-book)
1. Human Genome Project. 2. Human gene mapping.
3. Genetic engineering—Moral and ethical aspects.
4. Genomics—Moral and ethical aspects. I. Richardson,
Sarah S., 1980– II. Stevens, Hallam.
QH445.2.P678 2015
611′.0181663—dc23
2014044249

Permissions:

Chapter 4, "The Polygenomic Organism," by John
Dupré, first appeared in *The Sociological Review* 58
(2010) and is printed by permission of the publisher, John Wiley and Sons. © 2010 The Author.
Editorial organization © 2010 The Editorial Board
of *The Sociological Review*.

Parts of Chapter 8, "From Behavior Genetics to
Postgenomics" by Aaron Panofsky, first appeared
as part of his book, *Misbehaving Science: Controversies and the Development of Behavior Genetics*.
These parts are reproduced by permission of the
publisher, University of Chicago Press © 2014.
All rights reserved.

Cover art: Connectogram image courtesy of Dr.
John Darrell Van Horn, University of Southern
California.

CONTENTS

FOREWORD Biology's Love Affair with the Genome

Postgenomics is the unavoidable consequence of an intense love affair between biomedical scientists and the human genome. The discovery of the double-helical structure of DNA in 1953 lit the flame. The breathtaking rapidity with which this discovery lead to the entrenchment of the central dogma (DNA → RNA → protein), the cracking of the genetic code, the emergence of genetic engineering technology, and the early understanding of Mendelian diseases created an expectation of exponential increases in our ability to measure and interpret DNA information. DNA satisfies the compulsions of many scientists: measurable, discrete, molecular (yet apparently integrative), deterministic, and evolvable. If a little DNA sequence was good, then a lot—the genome—would be great. With the prospect of greatness, reasonable people are prone to hyperbole: save money, develop cures, predict disease, learn about our ancestors, and bring justice to all. It can be hard to judge harshly someone in love.

But even the most intense love affairs simmer and require nurturing. The breakneck speed of the courtship slows to a more reasoned set of discussions, negotiations, and settings of expectation. Some love affairs do not survive these adjustments, but others transition to a lifelong shared adventure. Postgenomics commences with an inventory of the successes and disappointments of the genome; we lift our heads, look around, and figure out what the future holds.

I do genomics, and I plan to do postgenomics. But this volume compels me to examine what I do and why I do it. The chapter authors combine a deep understanding of the history and technical content of modern genomic science with largely contrarian (to many genomicists, at least) interpretations of the significance and impact of the work. They expose significant biases in the way we formulate, justify, communicate, and defend work in genomics. Surprisingly, however, their analyses do not lead to despair, but to opportunity.

The chapters in this book highlight unrecognized and unexamined assumptions and suggest novel analyses and experiments. Evelyn Fox Keller refers to the "linguistic habits of geneticists"—habits that I have tried to master and also tried to avoid being fooled by. Keller reminds us that the genome's program is dynamic and reactive. John Dupré emphasizes that individuals are in fact combinations of multiple genomes. It is sometimes easy to overlook inconvenient facts that violate our abstractions, but as Mike Fortun celebrates in his marvelous meditation on "Toll!," an openness to the genome's surprises can gratify and motivate.

Adrian Mackenzie shows how as the genome's "shape" moves from linear to high dimensional, it provides more features than anyone can possibly interpret. In light of this, we may need to move to alternative representations of physiology that are more integrative. But, as Hallam Stevens makes clear in his analysis of network metaphors in postgenomics, the distinction between reductionist and holist is more complex than we thought. Epigenetics, for example, is thought to provide a nonreductionist mechanism for studying the interaction between gene regulation and the environment. As Sarah Richardson argues in her examination of maternal-fetal epigenetics, however, epigenetics claims often mirror classic genetic reductionist explanations in their focus on the mechanism of regulation of gene expression. Similarly, Sara Shostak and Margot Moinester point out that looking at environmental measures at multiple levels (molecular, cellular, tissue, organism) involves forms of reductionism that inevitably obscure some dimensions of the environment, including high-level environmental abstractions such as a neighborhood.

Intriguing questions about the reward and funding structure of the sciences accompany the postgenomic moment. One example is the increasingly dispersed nature of scientific knowledge production. As we integrate multiple databases, the legitimate coauthorship claims of data curators suffer from their distance (in both place and time) from the other authors. Rachel A. Ankeny and Sabina Leonelli explore how we can give credit to those who have shared and annotated data. Funders such as the National

Institutes of Health greatly benefit genome research, but funding focused principally on genomics can lead to distortions. Aaron Panofsky examines how certain areas of behavioral genetics have been "lavishly rewarded despite consistent failure to deliver," while Catherine Bliss looks at how genomics research in the field of race- and ethnicity-based health disparities may be crowding out public health and social science approaches.

Postgenomics: Perspectives on Biology after the Genome delivers important scientific and social messages. One scientific message is that the genome sequencing projects were neither unmitigated successes nor failures, but rather the start of a newly enabled era in which determining the sequence of four DNA bases is easy, but understanding its role in biological systems is incredibly challenging. One social message is that postgenomics should not be simply the playground of former genomicists now turned postgenomicists. Instead, there is a credible argument for a "reset" and evaluation of what the most promising and fruitful areas of investigation are likely to be. We should resist the temptation to merely declare the "obvious" next steps: epigenetics, environmental characterization, and large-scale population sequencing. Rather, we should pause and consider the range of societal and scientific responses to the past fifteen years of work and choose questions and strategies that allow us to marry discovery and its beneficial applications.

RUSS ALTMAN
Palo Alto, California
January 2014

1

Beyond the Genome

Hallam Stevens and Sarah S. Richardson

In the celebrations surrounding the completion of the Human Genome Project (HGP), few anticipated the bewildering developments that were to come. Expectations that the follow-up work would involve routine puzzle solving gave way to a series of surprising biological revelations. Debates over fundamental problems became more controversial and unsettled. Some important questions—which in the immediate aftermath of the HGP seemed near a solution—now, a decade on, seem even more difficult and mysterious.

The genome sequencing projects promised a future in which human traits would be linked to common genomic differences. Beginning in 2005, biologists began genotyping thousands of individuals, searching for correlations between single nucleotide polymorphisms and phenotypic traits. This technique for probing the meaning and function of the genome is known as the genome-wide association study (GWAS). To study obesity, for example, GWAS sampled thousands of obese individuals and thousands of nonobese individuals. If a particular mutation at location x occurred in a high fraction of obese people and a relatively low fraction of nonobese people, this suggested that location x might have something to do with obesity. This analysis was repeated for millions of locations on the genome, building up an overall picture of locations associated with particular traits.

By 2010, over seven hundred GWAS had been published on over four hundred different diseases and traits.[1] As more and more studies were conducted, two trends began to appear. First, many traits—even traits that biologists might have supposed to be quite straightforward—turned out to be associated with hundreds or even thousands of locations on the genome. One 2010 study associated 180 distinct locations with human height.[2] Second, even with all these locations taken together, the numbers just did not add up. With height, for instance, studies of monozygotic twins suggested that 80–90 percent of the variation in human height is heritable. However, using GWAS to measure the contribution of each location to the overall variability in height showed that the contribution of each location to the overall variation was very small. Adding up all the contributions of the hundreds of locations only accounted for about 13 percent of the overall variation in human height.[3]

This persistent question of "missing heritability" has continued to dog genome research. Despite the many locations on the genome responsible for influencing particular traits or diseases, there does not seem to be "enough" to actually account for what is going on. Biologists have put forward numerous suggestions to explain what might be going wrong: rare variants, copy-number variations, network effects, environmental effects, and epigenetic effects. The evolutionary biologist Leonid Kruglyak argues that the problem is conceptual: "It's a possibility that there's something we just don't understand, that is so different from what we're thinking about that we're not thinking about it yet."[4] Attempts to solve this problem have been a major motivation for studies of epigenetics and gene-environment interactions, as well as for projects to sequence large numbers of complete genomes (such as the 1000 Genomes Project). The missing heritability problem suggested that the working of the genome was far more complex than biologists hoped or expected it to be: traits and diseases seemed to depend on a mysterious set of unknown unknowns. This indicated the need not only for new experimental tools but also for a fundamental rethinking of the working of genes and genomes.

GWAS constitute only one of the promising developments within biology over the past decade. Yet, the kinds of problems they have encountered—more data, increasing complexity, greater uncertainties—are exemplary. Today, "postgenomics" is an increasingly prevalent term within the life sciences. Dozens of recent texts in bioinformatics, genetics, and medicine promise to situate research in these fields in "the postgenomic age."[5] Biologists invoke the term "postgenomics" to signal powerful new methods and approaches to complex biological problems. Similarly, social science analysts of the life sciences use the term "postgenomic" to refer to widespread transformations said to be sweeping these fields since the completion of the major genome projects. A recent editorial in the *Economist* titled "Biology 2.0" predicted that "It seems quite likely that future historians of science will divide biology into the pre- and post-genomic eras."[6]

This book aims to reflect on the postgenomic moment by posing a set of critical questions: What are the continuities and discontinuities between the postgenomic life sciences and previous biology? Which of the hopes and ambitions of the genome projects have been realized, and which have not? How does postgenomics transform fundamental conceptual debates in the life sciences, such as those over holistic versus genetic determinist approaches, biological and socio-environmental explanations, and how to conceptualize human racial and sexual differences? And how should we characterize the relationship between new high-throughput data technologies and new modes of postgenomic investigation?

Biology has come a very long way from the iconic 1992 moment when Walter Gilbert waved around a compact disk and asserted that the As, Ts, Cs, and Gs of the DNA sequences encoded on it *were us*. As the essays in this book demonstrate, science studies scholars examining the life sciences, too, are reimagining their role and refreshing their theoretical toolkits in the postgenomic age.

This book explores postgenomics as a live and evolving frame for discussions of changes and trends in the post-HGP life sciences. Following Richardson, we define postgenomics both temporally, as the period after the completion of the sequencing of the human genome, and technically, in reference to the advent of whole-genome technologies as a shared platform for biological research across many fields and social arenas.[7] We characterize as "postgenomic" all of those areas of the biological and medical sciences that now use genomic information or approaches as a foundational or standard element of their research practices. "Whole-genome" technologies include human genome databases and biobanks; microarray chips for assessing the expression of hundreds of thousands of genes in human tissue; rapid, inexpensive whole-genome sequencing technologies; high-throughput screening techniques; bioinformatic and computational advances in GWAS; and low-overhead, mail-order mass sequencing and genome analysis facilities.

For many, postgenomics signals a break from the gene-centrism and genetic reductionism of the genomic age. As scientists narrate the history of the genome sequencing projects, they trace a path from a simplistic, deterministic, and atomistic understanding of the relationship between genes and human characters toward, in the postgenomic era, an emphasis on complexity, indeterminacy, and gene-environment interactions. In this sense, "epigenetics," the study of mechanisms that regulate gene expression in response to environmental signals, is an archetypal postgenomic science. A 2007 NOVA *Science* feature presented epigenetics as a groundbreaking paradigm shift in biology and medicine.[8] Epigenetics, some claim, represents the new age of genomics in which nature and nurture are seen to interact in profound ways that overturn the old reductionism and determinisms of Watson and Crick's genetic code. The popular science press trumpets that epigenetics shows that "it's not all in the genes" and even that "you can change your genes." Public health activists have championed epigenetics as a mechanism for improving human health through environmental and sociostructural changes. In evolutionary biology, some have argued that epigenetics provides a framework for revisiting Lamarckian ideas, claiming that contrary to gene-centric dogma, life experience, encoded in the epigenome, can be passed on to future generations.

Yet not all are prepared to accept such pronouncements. In a widely circulated 2010 essay, *Current Biology* editor Florian Maderspacher offered a ringing critique of the popular and scientific frenzy over epigenetics.[9] "What is all the fuss about?" Maderspacher asked. The idea that genes are regulated by transcription factors has been well established since Jacob and Monod's classic *lac* operon experiments in the 1950s. Epigenetics is not new, nor is it revolutionary. Its draw today, Maderspacher asserted, is due more to our continued search for confirmation for our social ideologies in the facts of biology. Epigenetics has become fashionable, he argued, as a result of its politically liberal "sense of empowerment. . . . Much like with the idea of the vulgarizing genetic determinism . . . epigenetics seems to offer solid scientific proof—DNA modification as a kind of liberation."[10]

The critique provoked polarized responses, recorded in the comments section of *Current Biology*, that reveal the contested and emergent status of the young postgenomic science of epigenetics. Even as its wonders are celebrated in the popular press, many scientists, such as the evolutionary biologist Jerry Coyne, remain highly skeptical about epigenetics' novelty and about its implications for traditional models of inheritance and development.[11] Others, such as evolutionary biologists Marion Lamb and Eva Jablonka, insist that epigenetics presents a transformative critique of the central dogma and a mechanistic framework for the study of environmental factors in development and evolution.[12] Despite all the attention to epigenetics, its significance and even its meaning are highly contested.

The most recent exemplar of these multiplying contestations and difficulties is the Encyclopedia of DNA Elements (ENCODE) project, which announced its results in 2012. The ENCODE project began in 2003—just as the HGP was wrapping up—with the aim of identifying all functional elements in the human genome. ENCODE was one of the largest attempts to try to understand the three billion letters of the human genome. Initial analysis of the human genome had suggested that since less than 2 percent of the nucleotides coded for proteins, the rest was "junk DNA." The project set out with the goal of building a far more comprehensive "parts list" that included non-protein-coding genes, transcriptional regulatory elements, and sequences that mediated chromosome structure and dynamics.

When the ENCODE Consortium announced their findings, they claimed that over 80 percent of human DNA had some function.[13] Many celebrated this remarkable finding, proclaiming the death of junk DNA and predicting that textbooks would have to be rewritten. The ENCODE results were widely reported in mainstream media in glowing terms. But some biologists greeted the 80 percent figure with disbelief and outrage. Taking first to the blogosphere and then to journals, the critics argued that "ENCODE accomplishes [its] aims mainly by playing fast and loose with the term 'function,' by divorcing genomic analysis from its evolutionary context and ignoring a century of population genetics theory, and by employing methods that consistently overestimate functionality."[14] The thrust of this objection was that comparative genomics analysis had consistently found no more than 10 percent of the human genome to be under active selection. How could, then, 80 percent of the human genome be functional? Could 70 percent of the genome be somehow immune to evolution? As the critics pointed out in scathing terms, ENCODE had chosen a definition of "function" that ignored evolution completely.[15]

Inevitably, "postgenomics" has different meanings for biologists in various disciplines and research settings, and in the dispersed and fast-paced world of genomic science the meaning of the term has the potential to evolve rapidly. Biologists also seek to *do* different things—including carve out aspirational futures and seek funding and investment—with the concept of postgenomics. Conceptions of postgenomics serve to mobilize new agendas in the open play of the twenty-first-century biological frontier. "Postgenomics" has taken root in a competitive research environment in which private and governmental funders are wondering whether to continue to invest in genomic sciences, technologies, and knowledge despite their well-advertised failures to deliver cost-effective and novel advancements with direct clinical applications. Postgenomics, it might be argued, is, among other things, a mode of extending the project of genomics in the face of a restricted resource environment and increasing skepticism about genomics' prospects to provide concrete benefits for society and for human health.

Some of ENCODE's critics have suggested that the 80 percent figure was a publicity-grabbing stunt designed to generate greater excitement about genomic approaches and procure more investment in this particular mode of high-throughput research. This is what Mike Fortun has termed the "promissory mode" of genomics: a reliance on continuously renewed and hyped, but rarely fulfilled, promises of providing therapeutic and technological breakthroughs.[16] Indeed, some of the discomfort about the ENCODE project was its "big science" flavor: a centrally coordinated project, involving over four hundred scientists, critics argued, was the wrong approach to answering the kinds of questions that ENCODE was asking. Postgenomics has provoked increasingly sharp disagreements about how to share (often diminishing) money and rewards among institutions, projects, and individuals.

But the ENCODE controversy is not just about hype, methodology, and organization—it also shows how far biologists are from an agreement on an account of genomic action. It is not merely that the definitions of fundamental terms—such as "functionality"—are still open to contestation. The debate points to significant uncertainty about the overlapping roles (and relative importance) of evolution, DNA structure, transcription, and regulation in the human genome. Genes, the *New York Times* reported in 2008, are having an "identity crisis": acknowledgement of the importance of epigenetic marks, alternative splicing, post-transcriptional modification, and noncoding RNA have rendered the concept almost meaningless.[17] *Nature*'s "Genome at Ten" issue ran with the headline "Life Is Complicated": "The genome promised to lay bare the blueprint of human biology. That hasn't happened. . . . Instead, as sequencing and other new technologies spew forth data, the complexity of biology has seemed to grow by orders of magnitude."[18] The notion of the gene as "master molecule" is gone, but the disagreements exposed uncertainties about how else to talk about the proliferating objects, relationships, and levels involved between DNA and phenotypes. Rather than settling debates, ENCODE has muddied the waters; rather than answering older questions, it has raised new ones.

In contrast to the confident announcements that accompanied the HGP a decade ago, prominent biologists now caution against overoptimism. "The consequences for clinical medicine," Francis Collins said of genomics in 2011, "have thus far been modest."[19] Likewise, genomic entrepreneur Craig Venter suggests that there is still a long way to go to transform human genomes into medical knowledge: "The challenges facing researchers today are at least as daunting as those my colleagues and I faced a decade ago."[20]

More than a decade after the completion of the HGP, we stand at a moment of transition and contestation. During this "postgenomic era," genomes have become a shared platform for biological research across many fields. But as the examples given here suggest, fundamental tenets of the genomic research agenda have been challenged from all directions. Amid the deep controversies and disagreements within biology, what role exists for humanities and social science analysis?

Since the HGP was first proposed in the 1980s, science studies scholars have held a critical—and at times oppositional—posture toward genomics. They led the way in exposing the genetic reductionism and bald biological determinism of many genome sequencing boosters. They insisted on the need for social and ethical reflection on the potential harms of DNA technologies. This produced a vibrant decade of engaged and highly influential science studies scholarship on human genetics. Historians, philosophers, sociologists, and anthropologists worked intensively to understand the social, cultural, economic, and political consequences of genomics, examining the cultural transformations wrought by genomics around notions of identity, nation, affiliation, privacy, and personal health.[21] Despite their critical bent, social science scholars earned the respect of practicing scientists, finding allies and interested audiences among geneticists seeking to place their work in a wider context.

In the wake of the completion of the major genome sequencing projects, a more quiescent decade followed, as scientists, science studies scholars, and activists waited to see what the next big breakthroughs would be.[22] The essays in this book step into this open space to put the myriad changes instantiated within the life sciences in the post-HGP period in historical, social, and political context and thereby begin to make sense of them. Exploring the uncertain, transitional, and contested terrain of postgenomics, the contributors adopt a posture of critical reflection mixed with surprise and appreciation for the insights brought by the postgenomic era. They examine the continuities and ruptures, as well as the micro and macro transformations, instituted by the spread of genomic data and technologies throughout the life sciences over the past decade. They balance critiques of the practices and prospects of genome science with grounded analysis of the conceptual shifts and changes in everyday practices initiated by genomic data and technologies in knowledge production. Collectively, these essays document how postgenomics is reshaping debates over genetic determinism, reductionism, the role of the social and the environmental in human health and disease, and even the notion of the genome itself.

Notes

1. Stranger, Stahl, and Raj, "Progress and Promise of Genome-Wide Association Studies."

2. Lango Allen et al., "Hundreds of Variants."

3. For a useful overview of some of the criticisms of GWAS and some responses see Visscher et al., "Five Years of GWAS Discovery."

4. Quoted in Maher, "Personal Genomes."

5. Exemplary are titles such as Dardel and Képès, *Bioinformatics*; Giordano and Normanno, *Breast Cancer in the Post-genomic Era*; Plomin et al., *Behavioral Genetics in the Postgenomic Era*. The journal *Genome Biology* has refreshed its slogan, which now reads "Biology for the post-genomic era," and new journals are emerging to meet the interest in postgenomics, including the *Journal of Postgenomics: Drug and Biomarker Development* and the *Journal of Biochemistry and Molecular Biology in the Post Genomic Era*.

6. "Biology 2.0."

7. Richardson, "Race and IQ in the Postgenomic Age."

8. Holt, "Epigenetics."

9. Maderspacher, "Lysenko Rising."

10. Ibid.

11. Coyne, "Is 'Epigenetics' a Revolution in Evolution?"

12. Gissis and Jablonka, *Transformations of Lamarckism*.

13. National Human Genome Research Institute, "ENCODE Data Describes Function of Human Genome."

14. Graur et al., "On the Immortality of Television Sets."

15. Doolittle, "Is Junk DNA Bunk?"

16. Fortun, *Promising Genomics*.

17. Zimmer, "Now: The Rest of the Genome."

18. Check Hayden, "Human Genome at Ten."

19. Collins, "Has the Revolution Arrived?"

20. Venter, "Multiple Personal Genomes Await."

21. Of standout importance here are Duster, *Backdoor to Eugenics*; Fortun and Mendelsohn, *Practices of Human Genetics*; Nelkin and Lindee, *DNA Mystique*, 1st ed.; Kevles and Hood, *Code of Codes*; Keller, *Century of the Gene*.

22. Early explorations of postgenomics within biological theory and practice include Morange, "Post-genomics, between Reduction and Emergence"; Hans-Jörg Rheinberger, "Introduction to 'Writing Genomics'"; Stotz, Bostanci, and Griffiths, "Tracking the Shift to 'Postgenomics'"; Wynne, "Reflexing Complexity."

2

The Postgenomic Genome

Evelyn Fox Keller

In 1999, Francis Collins, director of the National Human Genome Research Institute, wrote, "The history of biology was forever altered a decade ago by the bold decision to launch a research program that would characterize in ultimate detail the complete set of genetic instructions of the human being."[1] His words were prophetic, though perhaps not quite in the ways he imagined.

The launching of the Human Genome Project (HGP) in 1990 initiated at least three turning points in the history of genetics: first and most extensively discussed in the wider literature is the role it played in the rise of "biocapitalism" and the commercialization of genetics; second is its role in the resurrection of a genetics of race; the third—my subject here—is the conceptual transformation its findings provoke in our understanding of genes, genomes, and genetics.[2]

Once the sequence of the human genome became available, it soon became evident that sequence information alone would not tell us "who we are," that sequence alone does not provide the "complete set of genetic instructions of the human being." Indeed, many have commented on the lessons of humility that achievement brought home to molecular biologists. The genome is not the organism.

But neither is it any longer a mere collection of genes. Barry Barnes and John Dupré have described the transformation effected by the new science of genomics as a shift that "involves genomes rather than genes being treated as real, and systems of interacting molecules rather than sets of discrete particles becoming the assumed underlying objects of research."[3] I want to argue that it has also changed the very meaning of the genome, turning it into an entity far richer, more complex, and more powerful— simultaneously both more and less—than the pregenomic genome. At the very least, new perceptions of the genome require us to rework our understanding of the relation among genes, genomes, and genetics. Somewhat more provocatively, I want also to argue that it has turned our understanding of the basic role of the genome on its head, transforming it from an executive suite of directorial instructions to an exquisitely sensitive and reactive system that enables cells to regulate gene expression in response to their immediate environment.

Throughout the history of classical genetics and early molecular biology, the science of genetics focused on genes, widely assumed to be the active agents that lead to the production of phenotypic traits. Similarly, the genome (a term originally introduced in 1920) was regarded as the full complement of an organism's genes. Indeed, I claim that this assumption is largely responsible for the widespread interpretation of the large amounts of noncoding DNA identified in the 1970s and 1980s as "junk DNA." Genomic research not only has made this interpretation untenable but also, I argue, supports a major transformation in our understanding of the genome—a shift from an earlier conception of the genome (the pregenomic genome) as an effectively static collection of active genes (separated by "junk" DNA) to that of a dynamic and reactive system (the postgenomic genome) dedicated to the regulation of protein-coding sequences of DNA. In short, it supports a new framework of genetic causality.

Of particular importance to this new perception of the genome were the early results of ENCODE, a project aimed at an exhaustive examination of the functional properties of genomic sequences.[4] These results definitively put to rest the assumption that non-protein-coding DNA is nonfunctional. By the latest count, only 1.2 percent of the DNA appears to be devoted to protein coding, while the rest of the genome is nonetheless pervasively transcribed, generating transcripts employed in complex levels of regulation heretofore unsuspected. The genome would now seem to be better viewed as a vast reactive system than as a collection of individual agents (embedded in a sea of effectively functionless DNA) that bear responsibility for phenotypic development.

In parallel with and in clear support of this transformation is the work of the past decade on the complexity and dynamism of genomic organization. Of particular recent interest is the relation, on the one hand, between the temporally specific conformational architecture of the genome and epigenetically regulated chromatin structure and, on the other hand, between conformation and gene expression. While it is certainly true that genome stability is required for reliable replication and segregation during cell division and meiosis, genomic plasticity appears to be equally essential. In particular, plasticity is required for the dynamic reorganization the genome routinely undergoes in the course of individual development. Tom Misteli writes that "the deceivingly simple question of how genomes function has become the Holy Grail of modern biology,"[5] and, indeed, it is precisely the pursuit of that question that now demands a reconception of the genome as an entity simultaneously stable, reactive, and plastic.

Genes, Genomes, and Genetics

Ever since the term "genome" was first introduced, it has been widely understood as the full ensemble of genes with which an organism is equipped. Even in the era of molecular biology, after the genome had been recast as the book of life, written in a script of nucleotides, it was not supposed that the instructions carried by the genome were uniformly distributed along the three billion bases of DNA. Rather, they were assumed to be concentrated in the units that "contain the basic information about how a human body carries out its duties from conception until death," that is, our genes.[6] To be sure, it has become notoriously difficult to fix the meaning of the term "gene," but, in practice, by far the most common usage that has prevailed since the beginnings of molecular biology is in reference to protein-coding sequences. Even if these protein-coding sequences make up only a relatively small fraction of our genomes, they continue to define the subject of genetics. As Collins wrote in 1999, these "80,000 or so human genes are scattered throughout the genome like stars in the galaxy, with genomic light-years of noncoding DNA in between." However vast, the "noncoding DNA in between" was not the object of interest. Until the early days of the new century, the primary focus of the HGP remained, as it had been from its inception, on the genes, on compiling a comprehensive catalog of protein-coding sequences.

Collins predicted that the full sequence of the first human genome would be completed by 2003, and he anticipated that it would produce a catalogue of roughly 80,000 genes. Other guesses ranged from 60,000 to 100,000.[7] His estimate of the completion date was right on target. But by that time, many of the expectations informing the launching of the HGP had already begun to unravel. The first jolt came in June 2000 with the announcement of the first draft of the human genome, reporting a dramatically lower number of genes (~30,000) than had been expected. Since then, the count has tended steadily downward, settling by 2003 at somewhere between 20,000 and 25,000—not very different from the number of genes in the lowly worm, *Caenorhabditis elegans*. Two questions became obvious: First, what, if not the number of genes, accounts for the vast increase in complexity between *C. elegans* and *Homo sapiens*? And second, what is the rest of the DNA for? Are we really justified in assuming that extragenic DNA makes no contribution to function?

The existence of a large amount of extragenic DNA was not exactly news, but its significance had been muted by the assumption that it was nonfunctional. When the HGP was first launched, it was widely assumed that extragenic DNA was "junk" and need not be taken into account. And indeed, it was not. Ten years later, a new metaphor began to make its appearance. Instead of "junk," extragenic DNA became the "dark matter of the genome," with the clear implication that its exploration promised discoveries that would revolutionize biology just as the study of the dark matter of the universe had revolutionized cosmology.

This shift in metaphor—from junk to dark matter—well captures the transformation in conceptual framework that is at the heart of my subject. It was neatly described in a 2003 article on "The Unseen Genome" in *Scientific American*, where the author, W. Wayt Gibbs, wrote, "Journals and conferences have been buzzing with new evidence that contradicts conventional notions that genes, those sections of DNA that encode proteins, are the sole mainspring of heredity and the complete blueprint for all life. Much as dark matter influences the fate of galaxies, dark parts of the genome exert control over the development and the distinctive traits of all organisms, from bacteria to humans. The genome is home to many more actors than just the protein-coding genes."[8] Of course, changes in conceptual frameworks do not occur overnight, nor do they proceed without controversy, and this case is no exception. The question of just how important non-protein-coding DNA is to development, evolution, or medical genetics remains under dispute. For biologists as for physicists, the term "dark matter" remains a placeholder for ignorance. Yet reports echoing, updating, and augmenting Gibbs's brief summary seem to be appearing in the literature with ever-increasing frequency.

To better understand this shift—or rather, to better understand the origins and tenacity of the framework that is now under challenge—I want to turn to a very brief account of the definition of the subject of genetics and the forging of its language back in the earliest days of the field. I argue that a number of assumptions that seemed plausible at that time were built into the linguistic habits of geneticists, and through the persistence of those habits, they have continued to shape our thinking about genetics to the present day—in effect, serving as vehicles of resistance to conceptual change. Primary among these have been the habits of thinking of genomes as nothing more than collections of genes and, as a logical consequence, of conflating genes with mutations.

The Language of Genetics

THE CLASSICAL AND EARLY MOLECULAR ERAS

For the paradigmatic school of T. H. Morgan, genetics was about tracking the transmission patterns of units called "genes." Even if no one could say what a gene was, it was assumed to be a unit, directly associated with a trait (a trait maker), and, at the same time, a unit that could also be associated with the appearance of a difference in that trait (a difference maker). When a difference was observed in some trait (a mutant), that phenotypic difference was taken to reflect a difference (a mutation) in the underlying entity associated with that trait—that is, the gene. Such an identification between phenotypic difference and underlying gene in fact required a two-step move. First, change in some underlying entity (the hypothetical gene) was inferred from the appearance of a difference in a trait (e.g., eye color, wing shape), and second, that change was used to infer the identity of the gene itself. Small wonder then that the first map of *Drosophila* mutations was so widely referred to as a map of genes.

Wilhelm Johannsen, the man who coined the term "gene," was himself aware of this problem, and it worried him. As early as 1923, he asked, "Is the whole of Mendelism perhaps nothing but an establishment of very many chromosomal irregularities, disturbances or diseases of enormously practical and theoretical importance but without deeper value for an understanding of the 'normal' constitution of natural biotypes?"[9] Clearly, genetics was meant to be more. Mapping "difference makers" and tracking their assortment through reproduction may have been all that the techniques of classical genetics allowed for, but the aims of this new field were larger. What made genes interesting in the first place was their presumed power to mold and to form an organism's traits. The process by which genes exerted their power in the development of characters or traits was referred to as gene action, and the hope was that the study of mutations would tell us how genes acted.[10]

In large part this hope was sustained by a chronic confounding of genes and mutations (and, at least by implication, of traits and trait differences)—a confounding that was built into classical genetics by its very logic. Genetics had come to be understood as the study of the units that were assumed to be strung together along the length of the chromosomes and that had the capacity to guide the formation of individual traits—in short, it was the study of genes. In 1920, an organism's complete set of chromosomes was named its genome. Of course, just what genes were, or what they did, was a complete mystery, but by the middle of the century, the "one gene–one enzyme" hypothesis of Beadle and Tatum seemed to provide at least a partial answer. What does a gene do? It makes an enzyme. Molecular biology soon provided the rest of the answer: it taught us that a gene is a sequence of DNA nucleotides coding for a sequence of amino acids, transcribed into a unit of messenger RNA, and translated into a protein.

Elegantly and seductively simple, this "purely structural theory" is of course too simple. For one, it leaves out the entire issue of gene regulation—that is, of the regulation of when, where, and how much of a protein is to be made. But Monod and Jacob were soon hailed for rectifying the omission, and they did so in a brilliant move that allowed for regulation without disrupting the basic picture. In short, they added "a new class of genetic elements, the regulator genes, which control the rate of synthesis of proteins, the structure of which is governed by other genes." In their operon model, such regulation was achieved through the presence of another gene (the regulator gene) coding for a protein that acts by repressing the transcription of the original structural gene. The discovery of regulator genes, they continued, "does not contradict the classical concept, but it does greatly widen the scope and interpretative value of genetic theory."[11] In short, even after the explicit incorporation of regulation, the genome could still be thought of as a collection of genes (protein-coding sequences), only now some of these genes were structural (i.e., responsible for the production of a protein that performs a structural role in the cell), while others did the work of regulating the structural genes. In this way, the central dogma holds, with genetic information still located in gene sequences, and the study of genetics still the study of genes.

JUNK DNA

In the early days of genetics, there would have been no reason to question the equation between genes and genetics. But from the 1970s on, especially as the focus of molecular genetics shifted to the study of eukaryotic organisms, and as the study of regulation assumed increasing centrality to that science, the relation between genes and genetics has become considerably less straightforward. To the extent that regulation is a property of DNA, it is surely genetic, but the question arises, is regulation always attributable to genes?[12] Clearly, the answer to this question depends on the meaning attributed to the word "gene," so let me rephrase: is regulation always attributable to protein-coding sequences?

A related (at least potential) challenge to the equivalence between genes and genetic material came from a series of discoveries of substantial expanses of non-protein-coding ("nongenic" or "extra") DNA sequences in eukaryotic genomes. Perhaps most important were the findings (1) of large amounts of repetitive DNA in 1968 and, later, of transposable elements; (2) of the wildly varying relationship between the amount of DNA in an organism and its complexity;[13] and (3) of split genes—protein-coding sequences interrupted by noncoding "introns" (identified in 1977). But the designation of such DNA as "junk"—a term first introduced by Susumu Ohno in 1972 to refer to the rampant degeneracy in eukaryotic DNA (i.e., to nongenic and, as he thought, nontranscribed sequences)—soon blunted that challenge.[14] After 1980, with the appearance of two extremely influential papers published back-to-back in *Nature*, the idea of "junk DNA" seemed to become entrenched.[15] These papers linked "junk" DNA to the notion of "selfish" DNA that had earlier been introduced by Richard Dawkins.[16]

Among other uses, the idea of selfish DNA was invoked to downplay the significance of transposable elements (in contrast to the importance McClintock attributed to them). The claim was that such elements "make no specific contribution to the phenotype."[17] But Orgel and Crick are explicit about their intention to use the term "selfish" DNA "in a wider sense, so that it can refer not only to obviously repetitive DNA but also to certain other DNA sequences which appear to have little or no function, such as much of the DNA in the introns of genes and parts of the DNA sequences between genes. . . . The conviction has been growing that much of this extra DNA is 'junk,' in other words, that it has little specificity and conveys little or no selective advantage to the organism."[18]

Until the early 1990s, the assumption that the large amounts of noncoding DNA found in eukaryotic organisms had "little or no function," that it contributed nothing to their phenotype and could therefore be ignored, remained relatively uncontested. For all practical purposes, genomes (or at least the interesting parts of genomes) could still be thought of as collections of genes. Indeed, when the HGP first announced its intention to sequence the entire human genome, much of the opposition to that proposal was premised on this assumption. Thus, for example, Bernard Davis of the Harvard Medical School complained that "blind sequencing of the genome can also lead to the discovery of new genes . . . , but this would not be an efficient process. On average, it would be necessary to plow through 1 to 2 million 'junk' bases before encountering an interesting sequence."[19] And in a similar vein, Robert Weinberg of MIT argued,

> The sticky issue arises at the next stage of the project, its second goal, which will involve determining the entire DNA sequence of our own genome and those of several others. Here one might indeed raise questions, as it is not obvious how useful most of this information will be to anyone. This issue arises because upwards of 95% of our genome contains sequence blocks that seem to carry little if any biological information. As is well known, our genome is riddled with vast numbers of repetitive sequences, pseudogenes, introns, and intergenic segments. These DNA segments all evolve rapidly, apparently because their sequence content has little or no effect on phenotype. Some of the sequence information contained in them may be of interest to those studying the recent and distant evolutionary precursors of our modern genome. But in large part, this vast genetic desert holds little promise of yielding many gems. As more and more genes are isolated and sequenced, the arguments that this junk DNA will yield great surprises become less and less persuasive.[20]

Weinberg was soon proven wrong. In the mid-1990s, and largely as a result of research enabled by large-scale DNA sequencing, confidence that noncoding DNA is nonfunctional, that it could be regarded as junk, began an ever more rapid decline. And by the dawn of the new century, few authors (not even Weinberg) still saw the equation between noncoding DNA and junk as viable.

THE EMERGENCE OF DARK MATTER: 1990 TO THE PRESENT

The launching of the HGP in 1990 was almost certainly the most significant moment in the entire history of our understanding of the term "genome." Although originally introduced as early as 1920, usage of the term remained rather limited before that time. After 1990, however, its usage exploded (figure 2.1).

But the rise of genomics did more than increase the term's usage; it also changed its meaning. It is possible to read the entire conceptual history of molecular biology as a history of attempts to integrate new findings attesting to unanticipated levels of biological complexity into its original basic framework, but I contend that with regard to my more specific focus on conceptualizations of the genome, the past two decades have been of particular importance. During this period, our view of the genome as simply a collection of genes has gradually given way to a growing appreciation of the

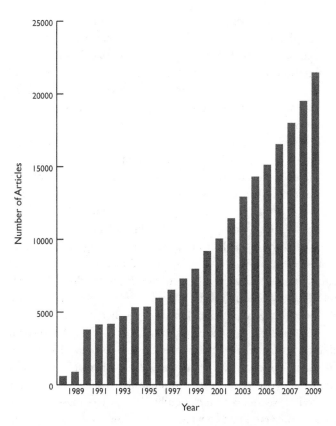

Figure 2.1. Number of articles referring to "genome" in biological literature, 1989–2010 (Web of Science).

dynamic complexity of the genome's architecture and its regulatory dynamics. Indeed, I suggest that, in the early years of the new century, genomic data brought earlier efforts of accommodation to a breaking point—a breaking point signaled by the shift in metaphor from "junk DNA" to that of "the dark matter of the genome."

Of particular shock value were the discoveries, first, of how few genes the human genome contained (discussed above) and, second, of how small a portion of the genome's structure is devoted to protein-coding sequences. In a review article published in 2004, John Mattick included a graph displaying the ratio of noncoding to total genomic DNA as a function of developmental complexity.[21] Prokaryotes have less than 25 percent noncoding DNA; simple eukaryotes have between 25 and 50 percent noncoding DNA; and more complex fungi, plants, and animals have more than 50 percent, rising to approximately 98.5 percent noncoding DNA in humans—who also have a genome size that is three orders of magnitude larger than that of prokaryotes (figure 2.2). Note that this analysis corrects for ploidy, whereas pregenomic estimations of the amount of DNA in different organisms did not. The different shading represent prokaryotes (bacteria and archaea; dark gray), simple eukaryotes (dots), *Neurospora crassa* (horizontal lines), plants (light gray), nonchordate invertebrates (nematodes, insects; diagonal lines), *Ciona intestinalis* (urochordate; hatches), and vertebrates (black); "ncDNA" refers to noncoding DNA, and "tgDNA" to total genomic DNA.

Mattick estimated the proportion of human DNA coding for proteins at 1.5 percent; since then, estimates have decreased to 1 percent.[22] The obvious question is, what is the remaining 98.5–99 percent for?

In 2003, the research consortium ENCODE (Encyclopedia of DNA Elements) was formed with the explicit mandate of addressing this question. More specifically, ENCODE was charged with the task of identifying all the functional elements in the human genome. Early results of that effort (based on the analysis of 1% of the genome) were reported in *Nature* in 2007, and they effectively put the kibosh on the hypothesis that noncoding DNA lacked function (i.e., that it was junk, "for" nothing but its own survival). They confirmed that the human genome is "pervasively transcribed" even where noncoding; that regulatory sequences of the resulting ncRNA may overlap protein-coding sequences, or that they may be far removed from coding sequences; and, finally, that noncoding sequences are often strongly conserved under evolution. Furthermore, they showed not only that noncoding DNA is extensively transcribed but also that the transcripts (now referred to as "noncoding RNA" or "ncRNA") are involved in many forms and levels of genetic regulation that had heretofore been unsuspected.

The reaction was swift. In his commentary accompanying the report, John Greally wrote,

> We usually think of the functional sequences in the genome solely in terms of genes, the sequences transcribed to messenger RNA to generate proteins. This perception is really the result of effective publicity by the genes, who take all of the credit. . . . They have even managed to have the entire DNA sequence referred to as the "genome," as if the collective importance of genes is all you need to know about the DNA in a cell. . . . Even this preliminary study reveals that the genome is much more than a mere vehicle for genes, and sheds light on the extensive molecular decision-making that takes place before a gene is expressed.[23]

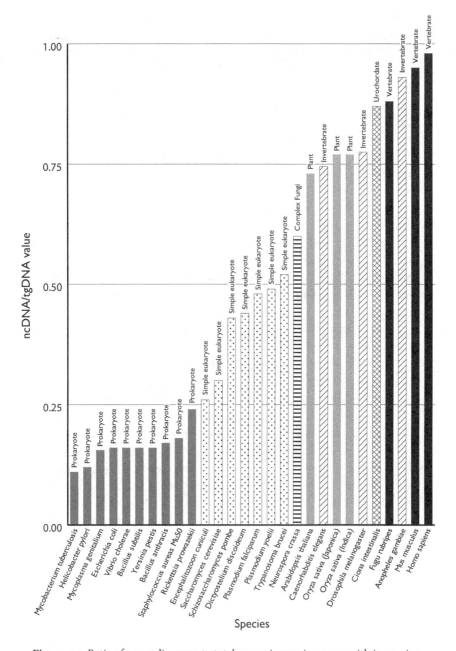

Figure 2.2. Ratio of noncoding DNA to total genomic DNA increases with increasing developmental complexity. Credit: Reprinted by permission of Federation of the European Biochemical Societies from Mattick, "The Central Role of RNA in Human Development and Cognition," figure 1.

Since 2007, efforts have been directed toward understanding just how the various kinds of ncRNA transcripts function in regulation. To this end, the ENCODE project has been expanded to include the genomes of a number of model organisms (e.g., *C. elegans* and *Drosophila melagonaster*), thereby making possible a comparative study of the relation between sequence and function.[24] The more complete results were finally released in 2012, and to much fanfare. They were accompanied by a special issue of *Nature* (September 6, 2012), a new publicly accessible website, and extensive coverage in the lay press. While some additional protein-coding genes have now been identified, the principle results focus on regulation. In an early summary of the new findings, Mark Blaxter identified three interacting systems that coordinate gene expression in space and time: "transcription factors that bind to DNA in promoters of genes, ncRNA that modifies gene expression posttranslationally, and marking of the histone proteins on which the DNA is wound with chemical tags to define regions of the genome that are active or silent."[25] Of particular interest are the strong correlations between chromatin marks and gene expression (apparently mediated by ncRNA) and the high degree of connectivity between and among different regulatory systems that have now been found in all the model organisms studied (also mediated by ncRNA).

The take-home message would seem to be clear. Genetics is not just about genes and what they code for. It is also about how the DNA sequences that give rise to proteins are transcribed, spliced, and translated into amino acid sequences, in the appropriate amounts at the appropriate time and place; about how these, once assembled into proteins, navigate or are transported to the sites where and when they are needed; and so on. All of this requires coordination of an order of complexity only now beginning to be appreciated. And it is now believed that the ncRNA transcripts of the remaining 98–99 percent of the genome are central to this process. These transcripts come in many sizes and are associated with a number of different mechanisms. Small RNAs can destabilize messenger RNA, influence the formation of chromatin and chromatin marks, and have even been linked to cancer. Now another class of ncRNA transcripts has been identified—"long intergenic noncoding RNAs" (or "lincRNAs")—that can operate across long distances and may prove as important to cell function as protein-coding sequences.[26] We have learned that ncRNAs are crucial to the regulation of transcription, alternative splicing, chromosome dynamics, epigenetic memory, and more. They are even implicated in the editing of other RNA transcripts and in modulating the configuration of the regulatory networks these transcripts form. In short, they provide the means by which gene expression can respond to both immediate and longer-range environmental contexts.

Implications

MEDICAL GENETICS

The implications for medical genetics, although slow in registering, have now begun to be taken to heart. For decades, it has been commonplace for medical geneticists to regard the significance of mutations in noncoding DNA exclusively in terms of their value in identifying the main actors of interest, that is, the genes responsible for disease. Thus, for example, the goal of the International HapMap Project (launched in 2003) is officially described as follows: "to develop a haplotype map of the human genome, the HapMap, which will describe the common patterns of human DNA sequence variation. The HapMap is expected to be a key resource for researchers to use to find genes affecting health, disease, and responses to drugs and environmental factors."[27] Elsewhere, the goal is described more succinctly: "to develop a public resource that will help researchers find genes that are associated with human health and disease."[28] Indeed, the confounding of "DNA sequence variation" with "genes that are associated with human health and disease" is reinforced by the explanations of this project routinely supplied in the official sources. These explanations make use of the term "allele," routinely defined as an alternative form of a gene that is located at a specific position on a specific chromosome. For example, on the home page of the HapMap website, we learn that "sites in the genome where the DNA sequences of many individuals differ by a single base are called single nucleotide polymorphisms (SNPs). For example, some people may have a chromosome with an A at a particular site where others have a chromosome with a G. Each form is called an allele."[29]

It is worth noting that this incorporation of SNPs into alleles (alternative forms of genes) is a fairly precise echo of the confounding of genes with mutations so common in the early days of classical genetics. It is similarly worth noting that the HapMap website also reiterates the assumption that SNPs will help locate and identify the relevant genes: "For geneticists, SNPs act as markers to locate genes in DNA sequences. Say that a spelling change in a gene increases the risk of suffering from high blood pressure, but researchers do not know where in our chromosomes that gene is located. They could compare the SNPs in people who have high blood pressure with the SNPs of people who do not. If a particular SNP is more common among people with hypertension, that SNP could be used as a pointer to locate and identify the gene involved in the disease."[30] Over the past couple of years, however, attention has begun to turn to other roles mutations in the DNA might have in disease, regulatory rather than structural. Examples of the role of mutations in micro and small ncRNA transcripts in the genesis of a wide range of diseases (e.g., cancer, Alzheimer's disease, cardiovascular disease, Tourette's syndrome, and a variety of developmental malformations) have been mounting steadily; similarly, but more recently, so too has the role of mutations in long ncRNA transcripts. In their recent review of the literature, Ryan J. Taft et al. cite leukemia, colon cancer, prostate cancer, breast cancer, hepatocellular carcinoma, psoriasis, ischemic heart disease, Alzheimer's disease, and spinocerebellar ataxia type 8 as examples of the many complex diseases in which lncRNA seems to play a primary role.[31] Furthermore, the authors go on to suggest that

> these examples likely represent the tip of a very big iceberg. The same technologies that have revealed a breadth of ncRNA expression are also driving a revolution in genome sequencing that will ultimately identify variations in the human genome that underpin disease susceptibility and aetiology. However, given the focus on mutations in protein-coding exons that cause most of the high-penetrance simple genetic disorders, the variation that occurs in non-protein-coding regions of the genome has, to date, largely been ignored or at least not been considered. This is changing: the emergence of genome-wide association studies to identify variant loci affecting complex diseases and traits and an increased awareness of ncRNA biology have prompted a reconsideration of the underlying protein-centric assumptions and provided a number of novel insights into disease-causing mechanisms.[32]

Genomic studies have also brought another kind of genomic anomaly to the attention of medical geneticists. Above and beyond the medical significance of local variations in ncDNA, there are also large-scale variations—differences in genomic architecture that have now been found to play important roles in a large number of complex diseases. "Genomic disorders" refers to a category of disorders that was first introduced by James R. Lupski in 1998 and that is associated with large-scale variation in DNA sequences (e.g., variation in gene copy number, genomic inversions, reciprocal translocations). As the author explained,

> Molecular medicine began with Pauling's seminal work, which recognized sickle-cell anemia as a molecular disease, and with Ingram's demonstration of a specific chemical difference between the hemoglobins of

normal and sickled human red blood cells. During the four decades that followed, investigations have focused on the gene—how mutations specifically alter DNA and how these changes affect the structure and expression of encoded proteins. Recently, however, the advances of the human genome project and the completion of total genome sequences for yeast and many bacterial species, have enabled investigators to view genetic information in the context of the entire genome. As a result, we recognize that the mechanisms for some genetic diseases are best understood at a genomic level.[33]

Ten years later, Lupski observes that "it is now becoming generally accepted that a significant amount of human genetic variation is due to structural changes of the genome rather than to base-pair changes in the DNA."[34] Indeed, the frequency of such structural changes appears to be orders of magnitude greater than that of single point mutations (SNPs), and Lupski's critique of traditional thinking is correspondingly strengthened: "Conceptually, locus-specific thinking had permeated genetics for over a century, with genocentric (gene-specific) views and base-pair changes as the one form of mutation predominating during the latter half of the 20th century and often blindly biasing genetic thinking to this day."[35]

EVOLUTION, ADAPTATION, AND THE POSTGENOMIC GENOME

A similar story has recently begun to unfold about the role of ncRNAs in evolution. Traditionally, for reasons partly technical and partly conceptual, the work of molecular evolutionary biologists has focused on changes in the protein-coding sequences of DNA. But with the growing recognition of functionality in ncDNA, that is beginning to change. As P. Michalak writes, "Noncoding RNAs control a remarkable range of biological pathways and processes, all with obvious fitness consequences."[36] Of particular excitement have been recent studies of the role of changing gene expression patterns in the evolution of primate cognition. In a recent study comparing gene expression patterns in human, chimpanzee, and macaque brains, M. Somel et al. report that much of the divergence between species—what distinguishes cognitive development in humans from that of other primates—is attributed to the changes in the expression patterns of genes that are regulated from afar by microRNA transcripts.[37] Relatedly, in a 2012 review of the roles of ncRNAs in brain evolution, Irfan Qureshi and Mark Mehler suggest that "because the proportion of noncoding DNA sequence present in the genome correlates with organismal complexity (in contrast to protein-coding genes), it is intriguing to speculate that ncRNAs preferentially mediated the accelerated and asymmetric evolution of the human CNS and underlie the unique functional repertoire of the brain."[38] Finally, Mattick and Mehler have also documented the extensive role of ncRNAs in editing RNA transcripts in brain cells, thereby providing a means for cognitive adaptation to changing environmental conditions. More specifically, they suggest that such editing provides a mechanism by which "environmentally induced changes in neural development and evolving brain architecture, cell identity and synaptic connectivity might subsequently be hardwired in the genome, potentially defining the complex and emergent properties of long-term memories and other structural and functional adaptations—a new paradigm and a plausible general molecular basis for dynamic and novel forms of learning and of potential mechanistic links between ontogeny and cognitive plasticity."[39] It is important to note, however, that behavioral (or physiological) adaptation does not require the direct alteration of DNA sequences: environmental signals trigger a wide range of signal transduction cascades that routinely lead to short-term adaptation. Moreover, by lending to such adaptations the possibility of intergenerational transmission, epigenetic memory works to extend short-term to long-term adaptation. As Mattick explains, "The ability to edit RNA . . . suggests that not only proteins but also—and perhaps more importantly—regulatory sequences can be modulated in response to external signals and that this information may feedback via RNA-directed chromatin modifications into epigenetic memory."[40]

Finally, environmental signals are not restricted to the simple physical and chemical stimuli that directly impinge: organisms with central nervous systems have receptors for forms of perception that are both more complex and longer range. Humans have especially sophisticated perceptual capacities, enabling them to respond to a wide range of complex visual, auditory, linguistic, and behavioral/emotional signals in their extended environment. Researchers have recently begun to show that responses to such signals can extend all the way down to the level of gene expression. In 2007, for example, Steve Cole and his colleagues compared the gene expression patterns in the leukocytes of those who felt socially isolated with the expression patterns of those who felt connected to others, and they were able to demonstrate systematic difference in the expression of roughly 1 percent of the genes assayed.[41] Subsequent studies have correlated gene expression patterns with other social indicators (e.g., socioeconomic status), providing further evidence for mechanisms of "social signal transduction" that reach all the way down to the level of the genome. Cole explains that "socio-environmental processes regulate human gene expression by activating central nervous system processes that subsequently influence hormone and neurotransmitter activity in the periphery of the body. Peripheral signaling molecules interact with cellular receptors to activate transcription factors."[42]

Conclusion: What Is a Genome Today?

The gap between a collection of protein-coding sequences and the full complement of genetic material (or DNA) of an organism is huge. Yet even so, and notwithstanding my earlier claims about the changes that have taken place in our understanding of the genome, that entity is still often regarded interchangeably as all of an organism's DNA, or as a collection of its genes, where the genes, the genome's constituent units, are tacitly taken to be effectively impervious to environmental input (see table 2.1).

Despite all the changes the gene concept has undergone, many of even the most recent formulations retain the view of these entities (and hence of genomes) as effectively autonomous formal agents, containing the blueprint for an organism's life—that is, all of the biological information needed to build and maintain a living organism. But I am claiming that current research in genomics leads to a different picture, and it does so by focusing attention on features that have been missing from our conceptual framework. In addition to providing information required for building and maintaining an organism, the genome also provides a vast amount of information enabling it to adapt and respond to the environment in which it finds itself, as indeed it must if the organism is to develop more or less normally and to survive more or less adequately.

I am proposing that today's genome, the postgenomic genome, looks more like an exquisitely sensitive reaction (or response) mechanism—a device for regulating the production of specific proteins in response to the constantly changing signals it receives from its environment—than it does the pregenomic picture of the genome as a collection of genes initiating causal chains leading to the formation of traits. The first job of the new genome is to detect the signals that impinge, and its second job is to respond (e.g., by change in its conformation) in ways that alter the patterns of gene expression. Eons of natural selection have ensured that the changes in gene expression patterns are appropriate to the new information. The signals impinging on the DNA come most immediately from its intracellular environment, but these, in turn, reflect input from the external environments of the cell and of the organism.

This reformulation gives rise to an obvious question: if the genome is so responsive to its environment, how is it that the developmental process is as reliable as it is? This is a question of major importance in biology, and it

Table 2.1. Definitions of "Genome" Obtained from a Range of Scientific Glossaries (Both Official and Semiofficial)

Source	Definition
Biology online (http://www.biology-online.org /dictionary)	The complete set of genes in an organism. The total genetic content in one set of chromosomes.
A Dictionary of Biology (Oxford University Press)	All the genes contained in a single set of chromosomes, i.e., in a haploid nucleus.
A Dictionary of Genetics (Oxford University Press)	In prokaryotes and eukaryotes, the total DNA in a single chromosome and in a haploid chromosome set (q.v.), respectively, or all the genes carried by this chromosome or chromosome set; in viruses, a single complement of DNA or RNA.
Glossary of genetic terms, Genetics Education Center, Univ. of Kansas Medical Center	All of the genes carried by a single gamete; the DNA content of an individual, which includes all fourty-four autosomes, two sex chromosomes, and the mitochondrial DNA.
Glossary, Human Genome Project Information, (http:// www.ornl.gov/sci/techresources /Human_Genome/glossary)	All the genetic material in the chromosomes of a particular organism; its size is generally given as a total number of base pairs.
National Human Genome Research Institute Glossary (http://ghr.nlm.nih.gov /glossary=genome)	The genome is the entire set of genetic instructions found in a cell. In humans, the genome consists of twenty-three pairs of chromosomes, found in the nucleus, as well as a small chromosome found in the cells' mitochondria. These chromosomes, taken together, contain approximately 3.1 billion bases of DNA sequence.
National Human Genome Research Institute, Genetics Home Reference Handbook (http://ghr.nlm.nih.gov/handbook /hgp/genome)	A genome is an organism's complete set of DNA, including all of its genes. Each genome contains all the information needed to build and maintain that organism.
Genome News Network, Craig Venter Institute (http://www.genomenewsnetwork .org/resources/whats_a_genome /Chp1_1_1.shtml)	A genome is all of a living thing's genetic material. It is the entire set of hereditary instructions for building, running, and maintaining an organism and passing life onto the next generation. The whole shebang.

Source	Definition
Genome News Network, Craig Venter Institute, Glossary (http://www.genomenewsnetwork.org/resources/glossary/index.php#g)	Gene: A piece of DNA used by cells to manufacture proteins, which carry out the business of cells. Each gene is a template for one or more proteins.
	Genome: A collection of genes. The human genome is the collection of human genes, just as the dog genome is the collection of dog genes. All living things have genomes. . . . A genome contains the biological information for building, running, and maintaining an organism—and for passing life onto the next generation. . . . A precise definition of genome is "all the DNA in a cell" because this includes not only genes but also DNA that is not part of a gene, or noncoding DNA.
National Center for Biotechnology Information, Science Primer, NIH (http://www.ncbi.nlm.nih.gov/About/primer/genetics_genome.html)	Life is specified by genomes. Every organism, including humans, has a genome that contains all of the biological information needed to build and maintain a living example of that organism. The biological information contained in a genome is encoded in deoxyribonucleic acid (DNA) and is divided into discrete units called genes.

E. F. Keller, "Genes, Genomes, Genomics," *Biological Theory* 6, no. 2 (2011): 132–40, table 1.

is rapidly becoming evident that the answer must be sought not only in the structural (sequence) stability of the genome but also in the relative constancy of the environmental inputs and, most importantly, in the dynamic stability of the system as a whole.[43] Genomes are responsive, but far from infinitely so; the range of possible responses is severely constrained, both by the organizational dynamics of the system in which they are embedded and by their own structure.

Changes in DNA sequences (mutations) clearly deserve the attention we give them: they endure, they are passed on from one generation to the next—in a word, they are inherited. Even if not themselves genes, they are genetic. Some of these mutations may affect protein sequences, but far more commonly, what they alter is the organism's capacity to respond effectively to the environment in which the DNA finds itself, or to respond differentially to altered environments. This conclusion may be especially important in medical genomics, where researchers routinely seek to correlate the occurrence of disease with sequence variations in the DNA. Since the sequences thus identified are rarely located within protein-coding regions of the DNA, the significance of such a correlation must lie elsewhere, namely, in the regulatory functions of the associated nongenic DNA.

Mutations also provide the raw material for natural selection. But when we speak of natural selection as having programmed the human genome, I want to emphasize that it is precisely the capacities to respond and adapt for which natural selection has programmed the human genome. Unfortunately, however, the easy slide from genetics to talk about genes, with all the causal attributes conventionally associated with those entities, makes this an exceedingly hard lesson to keep hold of.

Take, for example, the popular reports of a new study of the heritability of IQ, in which the distinction between genes and variations in nucleotide sequences simply disappears. The initial study claimed to "unequivocally confirm that a substantial proportion of individual differences in human intelligence is due to genetic variation,"[44] but it was popularly reported as showing that "50 percent of smarts stems from genes."[45] But that is not what it showed at all. Again, recalling that the great bulk of genetic variation occurs in regulatory regions of the DNA, we can see that the assumption that such variation refers to genes or gene causation is misguided. The far more likely consequence of genetic variation is an altered response to a particular environmental signal, a response that will persist through the generations as long as the relevant environmental signal persists.

A confirmation of this reading is provided by the surprising results of a different study—on an equally controversial subject—reported just three weeks later in the same journal. This time the focus is not on the heritability of IQ but on the heritability of spatial abilities, where the results of the reported study challenge one of the longest-standing assumptions of behavioral genetics. More specifically, the authors claim that "the gender gap in spatial abilities, measured by time to solve a puzzle, disappears when we move from a patrilineal society to an adjoining matrilineal society."[46] But if we accept the genome as primarily a mechanism for responding to our environments, this is hardly surprising. The only reason we had been misled to think of the correlation between sex and spatial ability as encoded in the genome is that we never imagined an environment so different from the usual as to elicit an entirely different correlation. It seems that sex may be correlated with spatial abilities, but only in patriarchal societies.

The bottom line: like other organisms, human beings are reactive systems on every level at which they are capable of interacting: cultural, interpersonal, cellular, and even genetic. The reconceptualization of the genome that I propose (from agentic to reactive) allows us—indeed obliges us—to abandon the dichotomies between genetics and environment, and between nature and culture, that have driven so much fruitless debate, for so many decades. If much of what the genome "does" is to respond to signals from its environment, then the bifurcation of developmental influences into the categories of genetic and environmental makes no sense. Similarly, if we understand the term "environment" as including cultural dynamics, neither does the division of biological from cultural factors. We have long understood that organisms interact with their environments, that interactions between genetics and environment, between biology and culture, are crucial to making us what we are. What research in genomics shows is that, at every level, biology itself is constituted by those interactions—even at the level of genetics.

Notes

Much of the argument in this essay is also discussed in Keller, "From Gene Action to Reactive Genomes," submitted after but published before submission of the present essay.

1. Collins, "Medical and Societal Consequences," 28.

2. See, e.g., N. Rose, *Politics of Life Itself*; Jackson, *Genealogy of a Gene*; Duster, "Race and Reification in Science."

3. Barnes and Dupré, *Genomes and What to Make of Them*, 8.

4. ENCODE Project Consortium, "Identification and Analysis of Functional Elements."

5. Misteli, "Beyond the Sequence," 796.

6. Collins, "Medical and Societal Consequences," 28.

7. At the annual genome meeting held at Cold Spring Harbor the following spring (May 2000), an informal contest was set up in which researchers tried to guess just how many protein-coding sequences it takes to make a human. As Elizabeth Pennisi wrote, "Some gene counters were insisting humans had upward of 100,000 genes, and just a handful were hinting that the number might be half that or fewer"; Pennisi, "Low Number Wins the GeneSweep Pool," 1484.

8. Gibbs, "Unseen Genome," 48.

9. Johannsen, "Some Remarks about Units in Heredity," 140.

10. For an overview of this history, see Keller, *Century of the Gene*, and *Mirage of a Space between Nature and Nurture*, chap. 2.

11. Monod and Jacob, "Teleonomic Mechanisms in Cellular Metabolism, Growth, and Differentiation," 394.

12. Although not directly germane to the main theme of this essay, it should be noted that a need to distinguish between genome-3 and genome-4 has also been

stimulated by current research—especially by the growing appreciation of alternative forms of inheritance (and hence of alternate meanings of "genetic")—that arises out of contemporary work on epigenetics.

13. This is generally referred to as the "C-value paradox"; see Thomas, "Genetic Organization of Chromosomes."

14. Ohno, "So Much 'Junk' DNA in Our Genome."

15. Doolittle and Sapienza, "Selfish Genes, the Phenotype Paradigm and Genome Evolution"; Orgel and Crick, "Selfish DNA."

16. Dawkins, *Selfish Gene.*

17. Orgel and Crick, "Selfish DNA," 605.

18. Ibid., 604.

19. B. Davis, "Human Genome and Other Initiatives."

20. Weinberg, "There Are Two Large Questions."

21. Mattick, "RNA Regulation," 317.

22. See, e.g., Blaxter, "Revealing the Dark Matter," 1758.

23. Greally, "Genomics," 783.

24. modENCODE Consortium et al., "Identification of Functional Elements."

25. Blaxter, "Revealing the Dark Matter," 1758.

26. Pennisi, "Shining a Light on the Genome's 'Dark Matter.'"

27. Available, as of January 7, 2014, at http://hapmap.ncbi.nlm.nih.gov /abouthapmap.html.

28. International HapMap Consortium, "Integrating Ethics and Science," 468; see also International HapMap Consortium, "International HapMap Project."

29. HapMap home page, http://hapmap.ncbi.nlm.nih.gov/abouthapmap.html, and repeated on the website of the National Human Genome Research Institute, "About the International HapMap Project," www.genome.gov/11511175.

30. "What Is the HapMap?," available as of January 7, 2014, http://hapmap.ncbi .nlm.nih.gov/whatishapmap.html.en.

31. Taft et al., "Non-coding RNAs," 131.

32. Ibid., 132.

33. Lupski, "Genomic Disorders," 417.

34. Lupski, "Genomic Disorders Ten Years On"; see also Stankiewicz and Lupski, "Structural Variation in the Human Genome."

35. Lupski, "Genomic Disorders Ten Years On."

36. Michalak, "RNA World," 1768.

37. Somel et al., "MicroRNA-Driven Developmental Remodeling."

38. Qureshi and Mehler, "Emerging Roles of Non-coding RNAs," 528.

39. Mattick and Mehler, "RNA Editing, DNA Recoding and the Evolution of Human Cognition," 231.

40. Mattick, "RNA as the Substrate for Epigenome-Environment Interactions," 551.

41. Cole et al., "Social Regulation of Gene Expression in Human Leukocytes."

42. Cole, "Social Regulation of Human Gene Expression," 133.

43. See, e.g., Keller, *Century of the Gene.*

44. Davies et al., "Genome-Wide Association Studies," 996.

45. Ghose, "Heritability of Intelligence."

46. Hoffman, Gneezy, and List, "Nurture Affects Gender Differences in Spatial Abilities," 14786.

3

What *Toll* Pursuit

AFFECTIVE ASSEMBLAGES IN GENOMICS AND POSTGENOMICS

Mike Fortun

Toll! Gate

In the late 1970s Christiane Nüsslein-Volhard and Eric Wieschaus, work-ing at the Max Planck Institute in Tübingen on the genetics of *Drosophila* development, developed their experimental system for producing tens of thousands of mutant fruit fly embryos.[1] As an "experimental system," their saturation screen using chemical mutagens produced not only novel "epis-temic things" but, equally important for this chapter, surprised scientists.[2]

One of the new "epistemic things" generated by Nüsslein-Volhard and Wieschaus was *toll*, a gene they showed to be involved in the control of dorsal-ventral polarity in the fruit fly embryo: *toll* mutants are all dorsal and no ventral, utterly one-sided. Nüsslein-Volhard has recalled gazing at this one of the thousands of mutant embryos that she and Wieschaus rou-tinely created and then examined under the microscope, when she uttered, "Das war ja toll!"[3] or perhaps simply "*Toll!*"[4] Accounts differ, but there is agreement on the need to provide an exclamation mark. This has been most frequently translated from German to English signaling pathways as "(That was) weird!" but "(That was) cool!" also has currency. This is the predominant meaning, positive or affirmative across its differences.

But all signaling pathways, linguistic and cellular alike, disseminate into multiple possibilities, increasingly variant. We have to add "Crazy!" to the *Toll!* pathway, as cued by Nüsslein-Volhard herself when, in an interview for a popular website, she reflected on a more metascientific level:

Interviewer: Hat es in Ihrem Leben den Heureka-Moment gegeben?
(Have you ever had a Eureka! moment in your life?)
Nüsslein-Volhard: Immer wieder mal. Das ist ganz toll! (Again and again.
That's what's so crazy!)[5]

Further semiotic branchings are suggested by FASEB *Journal* editor in chief Gerald Weissmann, who, in his own rumination on Nüsslein-Volhard's work and its affinities to Gestalt psychology, also invoked the complex network of meanings activated by *"Toll!"*: "Words shouted in the heat of discovery have more than their dictionary meaning. My emigré father used *'toll'* when he meant 'crazy,' but also 'curious' or 'amazing'; he used it when he first treated a patient with cortisone. These days German-speakers also use toll instead of 'cool' or 'droll,' 'outrageous' or 'awesome.'"[6] The entry of "droll" into our pathways here is rather *unheimlich* and reminds us of the capacity of any semiotic network to flip over into a near-opposite effect, just as the strange and the familiar suggest, imply, or produce each other.

I have digressed down these pathways early in this chapter so that these multiple effects of *"Toll!"*—predominantly positive, but intricately riddled with differential tensions—will remain in readers' memories. But I leave the conjoined translational instabilities of history and signaling pathways aside now and note here that Nüsslein-Volhard and Wieschaus's experimental system, innovative as it was, would almost certainly be described by professional historians of science as "genetic" and not "genomic," let alone as "postgenomic." No one at the time was talking about human genome projects, or genome projects for any other organism, for that matter. Although their overall project had a vaguely -omic intent—to catalogue, completely, all the developmentally important genes of *Drosophila*—historians would still say that their system worked in a genetic fashion, by creating and maintaining the lines that allowed for isolating the effects of particular mutant genes, characterizing them functionally or structurally at the molecular level, but with little or no use for or interest in DNA sequence.

Indeed, it was not until 1988 that Kathryn Anderson (a close colleague of Nüsslein-Volhard's) cloned *toll* and showed it to code for a transmembrane protein. It would be another five years before the next set of truly surprising results were produced. Between 1993 and 1996, Bruno Lemaitre and Jules Hoffman (among others) had established the key role of *toll* in a different signaling pathway, one that provides the fruit fly with protection against infection by *Aspergillus* and other fungi, as well as Gram-positive bacteria: a "developmental gene" had unexpectedly become an "innate immunity gene."[7] (How *toll*! is that?) But although the Human Genome

Project (HGP) had by then been debated and begun (debates that resulted in the inclusion of *Drosophila* and other "model organisms" in the project), and while DNA and protein sequence information had provided a number of interesting experimental clues and theoretical insights in the research linking development to immunity, much of this work could still be called "classical" genetics and biochemistry. In a 1996 article reviewing this research and discussing the evolutionary significance of the *toll* and other signaling pathways—pathways apparently developed before the plant–animal divergence as common protective mechanisms, and only later conscripted into developmental intricacies—Marcia Belvin and Kathryn Anderson never once use the words "genome" or "genomic."[8]

It was around that time that researchers in Charles Janeway's renowned immunology group at Yale cloned the first human gene coding for what would become known as a "*toll*-like receptor."[9] A key part of that work stemmed from searching the human gene sequence database at the National Center for Biotechnology Information—this was three years before even a "first draft" of a full human genome sequence would become available, that is, before "the Human Genome Project" is conventionally said to have been completed (in 2000). But if we periodized this work as "still genetic but becoming genomic," we would have to allow for its already becoming postgenomic as well.

Janeway's group had been expecting at the time to find a C-type lectin domain encoded in the clone they were browsing for. Ruslan Medzhitov later recalled the group's "initial disappointment" that their database search did not produce that result, but instead turned up a homology to a *Drosophila* gene.[10] It was only then that they learned about the "stunning discovery" of Lemaitre and Hoffman;[11] it seems that sometime in 1996 Hoffman showed Janeway a photograph (later the cover image of the September 1996 *Cell* issue, which bore their paper) of "a Toll-receptor deficient fly overgrown with aspergillus hyphae" at a grant meeting of the Human Frontiers in Science in which they collaborated. Janeway had his own "*Toll!*"-like moment.

Janeway's lab thereby initiated a new phase in the understanding of the mechanisms of innate (rather than adaptive) immune response in humans and popularized the new acronym for *toll*-like receptor, TLR, which "as an abbreviation" was "fast becoming as famous as RNA or DNA," evidenced by over 17,000 papers listed in PubMed in the following decade.[12] TLRs are what Janeway called "pattern recognition molecules" that are important nodes in the complex signaling pathways of cellular response, biosemiotic pathways that include numerous other proteins, as well as the more widely famed RNA and DNA. Such signaling pathways are hallmarks of "postgenomics," where agency and action are distributed rather than centralized in the gene, where codes become variably interpreted signals, and where an apparently immaterial "information" is always instantiated in material processes of "transduction."

I recount these brief stories not to demarcate genetic/genomic/postgenomic eras but to sidestep the historical demarcation problematic altogether. My intent is to pluck out from the immense flow of events in this time period, whatever we call it, a few episodes that highlight how scientific change is *affective* as much as it is cognitive, instrumental, experimental, and institutional. While the latter differences are indeed interesting and important to comprehend, the differences I am more interested in here may be summed up as differences in *"Toll!"*-like expression level. A significant change in the broader genetic-genomic-postgenomic history, of which the opening narrative above is a condensed and simplified metonym, is the increasing capacity of the different experimental systems to generate two different but entwined affects, surprise and interest—surprising, interesting, unprecedented objects like *toll*-like receptors, and surprised, interested scientific subjects who exclaim something like *"Toll!"* These affective assemblages will be difficult to assay, as we have limited methods and idioms for articulating these kinds of affective events. I also hope to develop, then, new idioms for the varied public spheres in which contemporary science is debated, evaluated, funded, and valued as a social resource, to better engage the kind of science that genomics has become— or perhaps continues to become—in postgenomics.

Stated somewhat differently, among the many signs we might use to differentiate genetics from genomics from postgenomics—immense and ever-expanding databases, with higher degrees of specificity and interconnection; ever-accelerating and increasingly frugal sequencing rates; proliferating robots, cheap and expensive; the changing amounts of time spent in "wet" versus "dry" activities; a plethora of job ads for bioinformaticians— there is a relatively neglected one: a complex affective response from scientists for which *"Toll!"* serves as a useful marker. What might we fish out of the historical archive of genomics and postgenomics using this marker, and might *we* in turn be surprised? And conversely, what might we seek to add to future historical archives of what-will-have-been-postgenomics that would further enrich those other signaling pathways, in which *the capacity for being surprised and interested* is a goal prized in the subjects of science, as well as the broader political and social worlds in which those subjects live and work?

Like genomicists mapping and sequencing a hitherto-unmarked stretch of chromosome, historiographers of genomics and postgenomics depend on certain tools and tropes to establish a sequence of events or other kinds of temporal, epistemic, institutional, technological, or symbolic difference in an otherwise yet-to-be-differentiated field of events. Like postgenomicists, historiographers who rely, as they must, on ready-to-hand (even if emergent and changing) tools will be prone to reductionism, oversimplification, and even "hype."[13] Like genomicists and postgenomicists, historiographers may find it productive to experiment with new tools and tropes that may seem overly speculative, insufficiently developed, and unwieldy or at least unfamiliar but may also generate fruitful new interpretive perspectives, that is, knowledge. This chapter, framed by the opening *toll*-like narrative above, suggests that attention to affect is worth experimenting with, worth risking failure with, and might add to our understanding of what is involved, or in play, in the ongoing development of postgenomics.

A Brief Note on Hypotheses: Surprise, Interest, and the Care of the Data

In the larger work of which this is a part, I hypothesize that there are deep connections between what is almost always troped as an overwhelming "avalanche," "flood," or "rush" of data found in contemporary postgenomics (and other technoscientific fields, often grouped under the rubric of "big data") and the creative and tacit "epistemic virtues" that I call "care of the data," the adroit, artful, and cautious handling of large data sets that permit both multiple interpretations and multiple errors, entangled.[14] These in turn are coupled with a palpable sense of excitement and eager anticipation that suffuse the inhabitants of the data-intense postgenomic science space. That sense of excitement, anticipation, and overall *toll*-ness is a forceful driver of scientists and sciences; however unaccountable, it deserves to be accounted for if we are to more fully understand why postgenomics (or any science) is so popular, fast, ubiquitous, and powerful.

Even a preliminary test of these hypotheses will be a long time coming, but it might incorporate current behavioral/environmental postgenomic explorations of how the startle response of zebrafish may be stunted by exposure to lead or other environmental toxins.[15] The capacity to be startled, at the neuro-physiological level, would seem to be an important part of our "inner fish," evolutionarily conserved and in complex interaction with an environment under various kinds of threat.[16] We could use the interesting results generated with that model organism work to scan human databases for promising homologous genes involved, before moving on to extensive (and expensive) genome-wide association studies comparing the distributions of alleles associated with startle behavior among postgenomicists to those of a control population of, say, historians. In early preparation for such (unlikely) work, here I begin only to briefly theorize this scientific subject as assembled from the fundamental affects that psychologist Sylvan Tomkins named "surprise-startle" and "interest-excitement."[17]

Tomkins described eight (sometimes nine) basic affects: interest-excitement and enjoyment-joy (the positive affects), surprise-startle (the only neutral affect), and the negative affects distress-anguish, anger-rage, fear-terror, shame-humiliation, and dissmell-disgust. The doubled terms were meant to convey the range of intensity in which the affect may be experienced. Affects are distinct from the more complex emotions that are co-assembled from the affects along with scripted cognitions. Emotions, in short, are what contemporary anthropologists would call "biocultural," while for Tomkins affects are a decidedly more physiological event, in the sense of being the property of a developing human organism as it encounters and responds, from birth, to a changing world.

"Surprise-startle" is the affect that I believe is embodied in the more complex, biocultural *Toll!*-like response that I have signaled with here; it "is ancillary to every other affect since it orients the individual to turn his attention away from one thing to another."[18] Surprise is "a general interrupter to ongoing activity"—the one-sided embryo in the long string of normals under the microscope, the sequence match from the publicly funded database that does not match the patterned expectation. I think it must be as fundamental to movement of science as it is to the human organisms who engage in it.

Later, I explore some "surprises" of the HGP, as recollected and expressed by a number of its leading figures, as part of a more encompassing argument that the *Toll!*-like response of surprise is the affect produced by any effective experimental system when it produces what Hans-Jörg Rheinberger terms the "unprecedented." Surprise-startle, for Tomkins, is a "circuit-breaker," a shatterer of habits that prompts a "re-orientation"—of the individual, but eventually of a community.[19] It is the response to what Evelyn Fox Keller calls "the funny thing that happened on the way to the Holy Grail"—the interruption of the expectation that "it might be a big, intricate code but it's still just code," or the realization that the code was always already broken or at least noisier and more open to multiple determinations than one dreamed, that every center and every command and every imagined mastery were decentered and disrupted and subject to deferred interpretive orders.[20]

With the burgeoning amounts of bio-semiotic-substance of all kinds produced daily by the investments in, of, and around the HGP, is it any wonder that funny things would happen—that interruptions and surprises would mount? Why were *we* surprised? Maybe we wouldn't have been so startled if we had learned to pay attention to the affective register of scientific developments.

"Interest-excitement" is the other related affect animating the postgenomic subject. Reoriented by the surprise-startle affect, a postgenomicist becomes interested in the accumulating information, which reinforces the interaction, driving the entire experimental system forward into new unprecedented futures. Like surprise, interest is elicited by something that could not be simpler—difference, as registered by "neural firing": "It is our belief that it is possible to account for excitement on a single principle—that of a range of optimal rates of increase of stimulation density. By density we mean the product of the intensity of neural firing times the number of firings per unit time."[21] We postgenomicists are interested by difference. The greater the difference—in intensity, or in number—the more neural activity, the more interest-excitement. And the more interest-excitement, the more one cares, and the more one cares, the better one thinks: "The interrelationships between the affect of interest and the functions of thought and memory are so extensive that absence of the affective support of interest would jeopardize intellectual development no less than destruction of brain tissue. To think, as to engage in any other human activity, one must care, one must be excited, must be continually rewarded. There is no human competence which can be achieved in the absence of a sustaining interest, and the development of cognitive competence is peculiarly vulnerable to anomie."[22] Excitement can be "massive," but it need not be; it is "capable of sufficiently graded, flexible innervation and combination to provide a motive matched to the most subtle cognitive capacities."[23]

In this extremely condensed version of my hypothesis, then, the gigascads of difference produced routinely in postgenomics is not just neutral information uploaded, downloaded, and regarded dispassionately; it excites and incites a desire for still more difference, faster. There are, of course, many more sources of interest-excitement in postgenomics than raw sequence difference, and the social, cultural, and cognitive contexts in which those differences are read (interpreted, differently) are enormously productive of scientific desire as well. I do not mean to disregard or discount those, but I am here exaggerating the more bodily, affective drivers of *Homo sapiens genomicus*, *Toll!*-kin of *Drosophila*, to remind us to account for affect when we are accounting for the many drivers of any "big data" science like postgenomics. That brightening of the eyes, that contraction of the facial muscles pulling the sides of the mouth upward into a growing grin that I have seen so many times on the face of a postgenomicist at the merest hint of a new data set or increased throughput rate, is not just an effect of sober rational assessment.

In the remainder of this chapter, I analyze some episodes in which startle-surprise and interest-excitement are drivers of, or are assembled along with, "subtle cognitive capacities" in some important developments in postgenomics—indeed, it may be a fairly good heuristic to think of postgenomics as what results when the comparatively coarse, crude, boorish, and "boring" genomics becomes somehow more subtle. One strand in this story is how DNA sequence information, as an object in itself, went from "boring" to *Toll!*

Avoid Boring Projects

Twenty years ago I wrote a history of the U.S. HGP as it was then emerging, focusing on the narrative structures and different kinds of work (scientific, social, political) scientists and others had to engage in to make genomics happen.[24] Taking the affective register seriously has required a revaluation of statements and events that I once overlooked or dismissed as unimportant in my previous writings on genomics. The main examples I take up here are expressions from the latter part of the 1980s, when the HGP was first being discussed and then initiated, concerning the "boring" quality of DNA sequencing as a form of scientific work and, somewhat less expressly, the boringness of DNA sequence data themselves, moving from there on to more recent events in postgenomics.

Let's begin with perhaps the most frequently read and referenced of statements regarding the "boringness" of DNA sequence and sequencing, from Walter Gilbert's 1991 *Nature* opinion piece, "Toward a Paradigm Shift in Biology." The opening paragraph conjured a mood, albeit a somewhat complex one: "There is a malaise in biology. The growing excitement about the genome project is marred by a worry that something is wrong—a tension in the minds of many biologists reflected in the frequent declaration that sequencing is boring. And yet everyone is sequencing. What can be happening? Our paradigm is changing."[25]

In his own effort to historicize his then-current location, Gilbert provided an important reminder to subsequent historiographers grappling with similar questions of change: if we are to map affects and their shifts, we will always be presented with complex amalgams. Malaise will be mixed with excitement, excitement will be marred by worry, and any of these component affects will be growing or diminishing, becoming more or less intense, but in any case always in some kind of tension with other elements.

So something was changing already in 1991, provoking the scientist to ask the historian's question: what can be happening?[26] "And yet" (there is always the "and yet" that marks the simultaneity of difference if not contradiction), sequencing is boring, "and yet" everyone is sequencing. There is excitement, "and yet" there is worry, boredom, malaise. Genomics is *Toll!*, and yet it is *Toll!*

Maybe Gilbert's statement simply begs for differentiation and specificity: *some* biologists in 1991 were bored with genomics, *some* were excited by it, and these different subject positions and their affects correlate with their social position: Nobel laureates, full professors at elite institutions, and (ex-)CEOs of successful biotech corporations may have enjoyed more freedom and privilege to find more excitement in genomic speculation than the assistant professor, postdoc, or technician worrying about the tedious and repetitive demands of grant writing, marker characterization and development, and quotas of DNA sequence to be fulfilled. Your *Toll!*-age may vary.

Some such social analysis is certainly possible and necessary, as will be seen below. But we should not explain away too quickly the force and logics of complex affective states and their possible importance for understanding scientific change over time.

Having used affect and collective mood to set the historical stage and open his essay, Gilbert never again invoked it. Affect faded into the background and became invisible, giving way to an equally complex amalgam of arguments about the reigning biological paradigm and the one Gilbert glimpsed emerging from it. This complex amalgam of arguments was framed by a statement that is neither inside nor outside Gilbert's text, an exclamation that appeared only in the subtitle and emanated from an editorial rather than authorial position: "The steady conversion of new techniques into purchasable kits and the accumulation of nucleotide sequence data in the electronic data banks leads one practitioner to cry, 'Molecular biology is dead—Long live molecular biology!'"[27] Gilbert's "own" trope for paradigm change (that is, the one that clearly occurs inside his authored text) was more definitive in its invocation of a "break," and more Chinese than French in historical connotation: "the view that the genome project is breaking the rice bowl of the individual biologist," he suggested, was a view that people had to get over.[28]

Despite his attunement to the revolutionary potential of genomic kits and data, Gilbert did not envision increased surprise, excitement, or similar *Toll!*-like affect as one of the outcomes characterizing the new era. "The tenfold increase in the amount of information in the databases will divide the world into haves and have-nots," Gilbert predicted, "unless each of us connects to that information and learns how to sift through it *for the parts we need*."[29] Genomics or postgenomics is, in this view, predicated on fulfilling what we already know we need and want, just faster and more efficiently. There is not really a sense that the *needs* themselves could, should, or would be transformed.

"Sequencing is boring" was indeed a frequent declaration in the debates in the mid- and late 1980s leading to the institutionalization of the HGP. James Watson's personal aversion to anything boring, especially boring people, has been well broadcasted.[30] Still, his concerns about boringness are worth taking into account, reflective as they are of broader cultural patterns. Early in his brief directorship of the HGP, Watson spoke in 1990 (around the time of Gilbert's article) about the problem of non-excitement at the American Academy of Arts and Sciences in Cambridge, Massachusetts:

> The people who wanted to do it [the HGP] were all old and almost retired, and everyone young was against it, because they figured if we did it, it would take money away from their research. So all the people you normally would expect, because they're going to do something, were against it, and all the people, you know, who really almost stopped [doing] science, were in favor of it. Now that includes me: I was really in favor of it, as was Paul Berg. And you could say that the objective was a wonderful objective. What's more important than this piece of instructions? But everyone else felt essentially frightened. It was going to be big science, it was going to be very boring—just determine all these letters—so anyone who would do it is someone you wouldn't really want to invite to dinner anyways.[31]

And it was not simply sequencing that was regarded as boring, in Watson's view; the development of high-resolution genetic maps provoked a similar affective response: "The trouble about getting these genetic maps, was that doing it was very boring, and in fact David Botstein had put in a grant application and had been turned down by NIH [National Institutes of Health]: it was too dull to be good science. But in fact it was a sort of tool that you really needed."[32]

Not only are affects always amalgams—here, it should be evident that Botstein, at least, found some excitement in the boring work of developing better genetic maps—but these affect amalgams are always assembled to epistemic objects and their larger cultural webs. This is part of the reason why affects tend to disappear from view: debates about *tools* and *big science*, for example, are well-recognized concerns of historians, sociologists, and philosophers of science and technology—and of scientists and engineers, too. So what does it really matter if Watson, Botstein, or any other scientist is bored, when more social and collective things are at stake?

I promise to return to such questions, which are anything but boring (to me, anyway). For now let us follow Watson's remarks, which prompted an interesting exchange with Matthew Meselson. Meselson began, "Jim, in *Drosophila*, I don't know of a single gene that has been gone after intelligently, that hasn't been cloned with a little effort, even though we don't have the complete sequence of *Drosophila*. So I gather that with humans, it's different, because we can't do genetic crosses and certain other manipulations as well with humans—but they might come along. So I would like to hear you explain why [this project is so] necessary for humans." Watson's initial response ("I think it's necessary in *Drosophila*, just because to get them all to cost—if we can do it at one-tenth the cost that it's being done in your lab, eventually it will be cost-effective. It won't be cost-effective if you do it at five to ten dollars a base pair, but if you do it at fifty cents a base pair, you'll get it out.") was hardly satisfactory to Meselson. "That's a different reason," Meselson argued back, one having little to do with doing science "intelligently"—that is, not by boring rote brainless mechanical means—so he restated his objection:

> *Meselson*: Not a single important gene that anyone has gone after intelligently has failed to be cloned and sequenced.
> *Watson*: Yeah, but there is a lot until we do it that you don't know the existence of, and the question is if you actually see the total thing, will you be surprised and get interesting scientific insights? And my guess is you will, but that's my . . .
> *Meselson*: That's a different reason than the one you gave. That's a good reason.

Unlike Gilbert, for Meselson productivity and efficiency are not particularly good reasons for dedicating $3 billion in public monies to something like the HGP—but to "be surprised" by "the whole thing" and get new "interesting scientific insights"? *That's a good reason.*

Now "good reason" may mean only a reason shared widely (enough) in the scientific culture of which Meselson and Watson are (elite) members— "Unexpected surprises? Sounds good to me!" But what if it *really is* a good reason—meaning, what if there were a shared understanding of a good society as one that contained, and cultivated, scientists wanting to be surprised and getting to be surprised, privileged as that position may be?

It is also worth remembering that even nonelites enjoy their affects, so I use my *Toll!*-like probe to pull up one more remembered event concerning these pre-postgenomic years. Going into my basement and accessing my dead-tree database, I found an interview I did as a pre-posthistorian of science graduate student with Robert Moyzis, then a leading scientist in the U.S. Department of Energy's (DOE) genomics programs. We had been talking—this was also in 1991—about the history of early meetings sponsored by DOE that are often credited as "precursors" to the HGP:

> *Fortun*: Can I just interrupt here for a second? It sounds from what you're telling me that the sort of standard account we get about mutation rates, and the technologies for detecting very small mutation rates, was almost on the side—or not as much of a concern or a goal as building resources.
>
> *Moyzis*: . . . I think in some ways that was an after the fact justification, in the sense that—I mean, keep in mind that DOE has always been historically interested in those problems . . . So that's not incorrect to say that that's what DOE's interests are, because that's always been DOE's interest and probably will remain DOE's interest. . . .
>
> [But] I think the history was more, "Hey, I'm real excited about this, this is a good idea; some aspects of this maybe are bigger science by biology's standards, therefore we're going to need organization, structure; average academic lab doesn't have that, DOE does." And then lastly, "Why the hell is DOE involved in this?" "Oh, well, you know, we want to study mutation, etc. etc." And I don't really think that that business about "gee, we want to understand mutation etc. etc.," really started happening until the criticism started. That when NIH started saying "what the hell is DOE doing this for? why is DOE interested?," that the sort of intellectual justification—which isn't an incorrect one. That does fit in with the mission, and I still believe that that's true, but you can't kid yourself that—I don't think it was that [sort] of a process. . . . [I]n fact in my mind, the kind of fruitcakes who kind of dreamt up this project, myself included, that's not where they were coming from . . . I mean, the genome, or DNA—it was kind of like this challenge to see if you could put the jigsaw puzzle together, as sort of an intellectual challenge. And many of the people at that Santa Fe meeting, I think, had that kind of mentality.
>
> . . . I think it's really after [the 1986 Cold Spring Harbor meeting]

that a lot of the apologizing and sort of re-writing of history to say, what's the scientific justification that the Department of Energy is involved in this, really began. So you've got a stretch there of perhaps almost months where, from my viewpoint, that issue wasn't even discussed. It was still, we can do this, it's worth doing, and it might actually be some fun doing, for a lot of crazy reasons. And I think it was certainly always discussed that there would be all these biomedical payoffs, but a lot of the initial players I don't think were even looking at it from that perspective.[33]

For Moyzis, the "intellectual justifications" for doing a human genome project, while clearly "not incorrect," were nevertheless secondary to "fun" and related "crazy [cool/awesome/exciting/*Toll!*] reasons." Promises of "biomedical payoffs" and other such rational justifications were certainly crucial to packaging, branding, and selling the HGP to its funders in the U.S. Congress, but it was the "intellectual challenge" of assembling a massive "puzzle" that causes it to be "dreamt up." (It is worth recalling that in the Kuhnian paradigm of scientific change, "puzzle-solving" is the mundane, perhaps boring work of "normal" nonrevolutionary scientists, not a challenge for fruitcake revolutionary ones.)

I have used these few episodes and recollections to characterize one aspect of resistance to a centrally organized effort to develop the sciences and technologies of genomics in the 1980s as predicated on its being "boring"—that is, not eliciting the affect of interest-excitement. Such resistance to the HGP was fairly widespread, with a far more complex quality than simply "it's boring," and with its own history, in which *some* resisters who were critical of *some* aspects of *some* of the project definitions that were put forward in the mid- to late 1980s came to be supporters of the HGP as its institutional and scientific definition emerged from various expert committees and bodies.[34] These affect threads were woven together with institutional turf politics and their attendant mix of scientific and ideological arguments about the shape and form of the HGP. In those expert discussions, which resulted in a less sequence-obsessed and more mapping-inclusive project, as well as the inclusion of various "model organism" genomes along with the human, the DOE and its scientists were often troped as rather mindless, good only for tool-building or engineering-type technological problems (sorting cells, building and shipping chromosome libraries, banking but not analyzing data, etc.), while the forces of the NIH were ones of creativity, able to pose and answer actual biological research questions. Because DOE=boring and NIH=interesting in these assessments, NIH would come to control twice as much money as DOE and be regarded as the "lead" agency. The affects associated with different sciences and scientists matter to sociopolitical events.

One of the most persistent strains of criticism at this formative time concerned the perceived threat that an expensive, centralized genomics program posed to individual R01 research grants, the mainstay of the NIH extramural program. Thus, scientists like Bernard Davis were asked to testify at a U.S. Senate hearing fairly late in the process, where U.S. senator Pete Domenici (Democratic senator from New Mexico [home of Los Alamos National Laboratory] and one of the key congressional advocates of the HGP) confessed that he was "thoroughly amazed . . . at how the biomedical community could oppose this project":

> I cannot believe that you are going to insist on business as usual in this field. It is beyond my comprehension, I repeat, beyond my comprehension. . . . [Y]ou cannot sit here and tell me that in all of the research that is going on with the marvelous individual investigators . . . you cannot tell me there is not more than $200 million, that if we even asked you to go look, you would say probably went for naught. . . . People had a lot of fun. Scientists had a lot of exciting mental activities. But it is inconceivable that out of $7 billion in grants in this very heralded R01 peer review approach . . . that there is not at least $200 million or $300 million that even one as dedicated to the field as you are could not go out there and look at and say, maybe we do not have to do this. . . . [There] are too many scientists in other fields that are using hardware and new technology and new techniques that support this as a tool, that I cannot believe that you really oppose it.[35]

The statement can be read as indirect confirmation that "fun," or the more clumsily phrased "exciting mental activities," is an important driver of scientific activity. Even senators, whose understanding of science (among other things) is notably limited, understand this. But Domenici also understood that playtime, in which "marvelous individual investigators" got paid with public dollars to try to surprise themselves—even if playtime was also its opposite, "business as usual"—was now over, and it was time to get on with the more serious, un-fun, and boring work of tool making, manufacturing the "hardware and new technology and new techniques" of the HGP.

Genomic Mandala

If the idea of loads of DNA sequence information in the mid-1980s elicited a *Toll!*-like affect in which "boring" predominated over "interesting," it did not seem to take long for that affect amalgam to invert its composition.

By the mid-1990s—after only a few years into the distributed, dedicated, federal-tax-revenue-supported, multiorganismal sequencing and mapping efforts shorthanded as the HGP—being hit with a flood of sequence information (and it indeed seems to have been dramatically physical experience for some) would elicit surprise and excitement, with barely a tinge of boredom.

An informative marker for this shift is the 1995 publication of the full sequence and map of *Haemophilus influenza*, signifying for some "the real launch of the genomic era."[36] Popular science writer Carl Zimmer noted that the publication of the 1.8 million base pair sequence landed "with a giant *thwomp*," disrupting what he wryly called "the dark ages of the twentieth century, when a scientist might spend a decade trying to decipher the sequence of a single gene."[37] Even though there were not "a lot of big surprises" about the microbe itself, Zimmer recalled that "what was remarkable was the simple fact that scientists could now sequence so much DNA in so little time." Moreover, that remarkable fact was transmitted in an instant, through the "kaleidoscopic wheel" mapping all 1,740 genes: "It had a hypnotizing effect, like a genomic mandala," reflected Zimmer, and "looking at it, you knew biology would never be the same."[38]

"And yet"—sameness always returns to level surprising difference; boredom reasserts its persistence in the interest-excitement affect amalgam. Even as genomicists "witnessed aspects of microbial diversity beyond what had been previously appreciated" in the years following the kaleidoscopic hit of the *H. influenza* genome, the mounting number of these "surprises" generated by the new sequencing capacities did not take long to simply become "widely accepted" features of the increasingly postgenomic landscape and lifeworld. Zimmer dubbed it the "Yet-Another-Genome Syndrome."[39] *Toll!*/cool had flipped to *Toll!*/droll.

Surprises of the HGP

And so, pointing out genomic surprises became something of a dull routine by the time the HGP had been ritually marked as completed in 2000. "It appears now that hardly a week passes without some new insight into the 'genome' taking us by surprise," noted biologist Richard Sternberg.[40] But despite their frequency, and the growing frequency with which they were thus noted, surprises still remained largely anomalous or background "color" to the main story.

Here I note a few of the other objects or events that elicited surprise and interest among genomicists; sequence information itself was an important but not the only part of these surprises. To its credit, when Cold Spring

Harbor introduced an oral history collection on "Genome Research" to its web pages, in addition to topical sections like "Mechanics of the HGP," "Challenges of the HGP," and "Gene Patenting," it also saw fit to ask its interviewees (all male on this topic), "What surprised you the most?" The affective register on display here is right on the surface—postgenomics turned out to be surprising, interesting, exciting, *Toll!*, anything but boring.

Perhaps the biggest and most publicized surprise was the number of human genes in the genome coming in much lower than expected. Sequencing innovator Bruce Roe, who led the group at the University of Oklahoma that first sequenced an entire human chromosome (22), remarked,

> [T]hat's an easy question. I put my dollar down on 120,000 genes. And for somebody to tell me that we only have twice as many genes as a worm, twice as many genes as a fly, you know, that's kind of disconcerting. . . . So that's one of my big surprises. The other big surprise was that there are genes overlapping genes. And genes inside of genes. Here we have this huge genome that only one and a half percent encodes for, and these genes overlap each other, you know. Why would you ever do that? Well, you know, I didn't design it. We're just looking at what the designer did.[41]

Robert Waterston mentioned another now well-known surprise of post/ genomics, "that fifty percent of the worm genes are shared with people, and fifty percent of human genes are shared with worm or something like that. It's just astounding."[42] The reorienting of evolutionary history and theory is a big part of my opening story of *toll*-like receptors, as researchers followed genes, gene functions, and gene sequences from fly to human and beyond, reorienting much of immunology in the process.

In this series of fireside chats (metaphorically and literally—these interviews were mostly shot over a few days in front of a Cold Spring Harbor fireplace), Eric Lander departed the most from sequence-centrism and waxed Weberian:

> What surprised me most? In the end, how satisfying it was personally. That I went into the project as a relatively young scientist. I began to get into the HGP at the age of thirty. At that point, you do things cause you're young and hotheaded and competitive and all sorts of things. Having now devoted fifteen years of my life to this, it's a very large piece of it and at some point I came to have—maybe about half way through—just a tremendous affection for the people, my colleagues doing it. A tremendous feeling like this was a purpose much greater than any of us, much bigger than me. It was the first time I felt like I was a part of something much

more important than I was and with a much greater purpose and something that would live far beyond me. That was why when the Celera thing came along and Craig [Venter] came along aiming to kill this, I probably reacted more strongly than anybody in the project, more violently in my reaction than anybody in the project because this mattered. It mattered to get right. It mattered because for me this was, you know, this was a calling in life and a purpose in life and nobody was going to go and screw it up like that and turn it into some private thing and not let us get the benefits from it. . . . I can't imagine a more wonderful thing to have done in life. And that surprised me. I guess I didn't ever imagine that it would end up meaning so much.[43]

Just to clarify, I admire that Lander waxed Weberian, and I would shout *Toll!* if there were more of a *Wissenschaft-als-Beruf* tone to our public discourse of science now. Maybe I just have a soft spot for Lander; in 1990 he was the first person I ever formally interviewed when I was a larval historian of science, and I was impressed then by his generosity, honesty, and insistence that the metaphor most appropriate to the HGP was not the "Holy Grail" but the less exciting, more boring public-infrastructure-oriented "Route One of Genetics." So in rereading this interview decades later I am surprised, in turn, that *he* was surprised that boring infrastructure would "end up meaning so much." Watching the video is a necessary supplement to reading the transcript; Lander pauses, appears to take the question quite seriously, ponders, and responds with spontaneity and enthusiasm.

Some of the most honest and reflective perspectives, and most resonant with our signaling pathways here, came from Maynard Olson:

There are a lot of things that surprised me about it. You know, history always looks so clear in retrospect, but not so clear in prospect done. I thought that it went much more rapidly and much more smoothly than I could have imagined. And it did so for a whole bunch of reasons. I mean, the problem was enormous. . . . Not conceptually difficult but practically an immensely difficult problem. . . . We were just mismatched. We didn't have good enough techniques. We didn't have anywhere near enough strong investigators. The whole computational infrastructure didn't exist. Most of our ideas about how to proceed were wrong. There was no overall organization that we had any kind of experience with. There was just the idea of building one. The problems just seemed rather overwhelming but, of course, exhilarating. I was never pessimistic although I was always restraining people that said, you know, this is going to be easy. . . . I think that if you had asked me how long it was going to take I would have been off by a decade or so. And even then I would have felt I was being optimistic. Because again I just couldn't see the path. And I couldn't see the path because the path wasn't there. A lot of other people thought that they could see the path but if you go back and read in detail what they said, that didn't turn out to be the path. Those were a lot of dead ends.[44]

So among the things genomicists were surprised by: genomicists were surprised by themselves.

In the conclusion, I return to this labyrinthine dimension of scientific work, in which the amazement generated in the doing of science is an affective effect of its mazelike qualities, and one is alternately overwhelmed and exhilarated (as with any encounter with a sublime). But next I consider how these brief, fragmentary glimpses of the affective dimension of postgenomics become difficult to account for, hard to assimilate into a system of historical or social value.

Insisting Genomics Account for Itself

Leapfrog again now to ten years after the ceremonial "completion" of the HGP. The *New York Times*, which had long exuded nothing much short of unalloyed enthusiasm for the HGP and every milestone discovery within that vast enterprise, conveyed a tone of disappointment that suddenly seemed to be everywhere in a 2010 editorial: "10 years later, a sobering realization has set in. Decoding the genome has led to stunning advances in scientific knowledge and DNA-processing technologies but it has done relatively little to improve medical treatments or human health."[45] If the "advances in scientific knowledge" and technology here said to be "stunning" were, in fact, stunning, it would be hard to imagine the sentence and the editorial moving on so quickly, in the same sentence, to that dismissive "but . . ." of accountability. I hardly expect the editorial board of the *New York Times* to be truly stunned by anything, including their own role in building enthusiasm for wreckless invasions that kill millions, but I wonder how an editorial written within a culture that had a greater capacity to convey and truly share in the scientist's sense of stunningness might read differently? Instead it's just: *stunning, yeah sure, but they haven't exactly cured anything like they promised to when they cashed that $3 billion check we gave them, have they?*

Nicholas Wade echoed the judgment and the rhetoric in an accompanying article: "For biologists, the genome has yielded one insightful surprise after another. But the primary goal of the $3 billion HGP—to ferret out the genetic roots of common diseases like cancer and Alzheimer's and then generate treatments—remains largely elusive. 'Genomics is a way to do science, not medicine,' said Harold Varmus, president of the Memorial Sloan-Kettering Cancer Center in New York, who in July will become the director of the National Cancer Institute."[46] It is as though surprise and insight were a dime a dozen, last year's news, and could not really disturb the calculations of worth that were being eagerly assessed. And it would do little good to point out the—irony? *Schadenfreude?*—that many historians, feminist philosophers, sociologists, and more than a few postgenomicists themselves had been making almost exactly that prediction about the expected medical benefits of the HGP for any number of years. *They* were not surprised when the relationship between genomes, illness, and medicine turned out to be more complicated than hoped or predicted by the more hard-core genome-as-Grail advocates.

What new genre of science writing could take account of the *Toll!*-like emergence—both and/or neither science and/or medicine, both and/or neither immunology and/or genomics—of the family of *toll*-like receptors and their fantastic evolutionary history that binds human to halibut, *Drosophila* to *Danio*, a history that includes the evolutionary repurposing (surely *that* is "the designer" that Bruce Roe invoked above?) of "genes for" innate immunity into "genes for" embryonic development? Public discourse on postgenomics—and similarly multiscale, distributed, complex, public-resource-dependent scientific projects, such as the sciences of climate change[47]—needs new genres of science writing that are patient with the difficult demands of interpretive multiplicity and openness, transparent toward and tolerant of complexities and ambiguities (marked here by my disseminated *Toll!* sign), yet resolute in their recognition and embrace of genuinely creative, *better* science.

Postgenomics has indeed gotten to be "better science" than genomics was: more subtle, less determinist, more attuned to the flexibilities and limits of its own categories, techniques, and analytic concepts.[48] Joan Fujimura and Ramya Rajagopalan's article on the theories and practices of genotype variation in large-scale populations ends with the claim (controversial to some social scientists who believe we should not issue "normative" judgments) that such practices and theories have, over time, become "better science"—scientifically better and ethically better, no longer supporting those unrealistic and harmful categories known as "race." "Better" for me also means "more careful," with that term again amalgamating the cognitive with the ethico-pragmatic with the affective.

In the larger project of which this chapter is a part, I am reorienting my positive affects toward a better ethnography of these better sciences, focusing on asthma researchers incorporating postgenomic findings and practices into a larger ecology of diverse sciences, from air quality modeling to psychosocial stress measurement to environmental and antipoverty activism. In that story, no gene ever came anywhere near to master-molecule magic-bullet holy-grail territory, yet postgenomicists have found much of interest. *Toll*-like receptors are one small part of that interest network, and like all the other multitudinous distributed parts, their signaling effects are always partial, sometimes contradictory, dispersed in unexpected ways between and within thoroughly mongrel populations, in play only at certain stages of development and under the sway of variable environmental conditions from the microbial flora of the gut to the ozone and particulate matter in the air of U.S. cities, maintained at deadlier-than-necessary levels to preserve the profits of energy corporations. Yet I admire the postgenomicists of asthma, even as their efforts and knowledge are swamped by the sublimity of asthma's causes and exacerbators. I admire that they, like so many other postgenomicists, have networked themselves into consortia that share data, results, materials, analytic techniques—and I admire as well their misgivings and uncertainties about all of these. And I particularly admire how they, like so many other postgenomicists, have transvalued even the "boring" work of curating and caring for tool-like materials/data sets, materials/data banks, and materials/data techniques so that these infrastructural activities have their own rewards, virtues, recognitions, and even surprises and interests far beyond what any dreamer of genomic futures had in the 1980s.[49]

The Amazement of Experimental Systems and the Public Sphere

To revisit and re-sound my opening story: "*Toll!*" the expression, like *toll* the gene, should register multiple effects in multiple signaling pathways, altered by multiple contingencies of context—environmental, temporal, developmental. "Weird!" and "Cool!" are only the first two dominant registers probed by the various translators of Nusslein-Volhard's speech act; "awesome," "crazy," "droll," and "mad" can also be activated.

Hence my preferred expression, as in my chapter's title echoing Francis Crick's autobiographical *What Mad Pursuit*: genomics became an increasingly *Toll!* pursuit—cool and awesome, but at the same time also a bit weird, boring, and more than a little mad.

These amalgams are essential: whereas genomics was *mostly felt to be* a boring machine tool, overcoded by the code discourse it directed at genomes to straightforwardly translate or decode them, postgenomics is *mostly felt to be* cool and awesome and crazy and weird and droll and mad, excitedly exceeding all its coded channels. Attending to affect should offer a complex of surprises and interests, often out of sync, not a simplified understanding of postgenomics.

"I couldn't see the path because the path wasn't there," Maynard Olson recalled about the paradoxically even-sooner-than-expected completion of the HGP. This view of the entire history of the HGP can be read as a confirmatory signal for Rheinberger's view of experimental systems as a maze:

> An experimental system can be compared to a labyrinth whose walls, in the course of being erected, simultaneously blind and guide the experimenter. The construction principle of a labyrinth consists in that the existing walls limit the space and the direction of the walls to be added. It cannot be planned. It forces one to move by means of checking out, of groping, of *tatônnement*. . . . The articulation, dislocation, and reorientation of an experimental system appears to be governed by a movement that has been described as a play of possibilities *(jeu des possibles)*. With Derrida, we might also speak of a "game" of difference. It is precisely the characteristic of "fall(ing) prey to its own work" that brings the scientific enterprise to what Derrida calls "the enterprise of deconstruction."[50]

In an age of austerity, it seems easy to forget, and easy to devalue, that postgenomics is a game that provides its players with a high degree of surprise and excitement—tempered by tedium. Why should we as a society pay (a lot!) to boost their levels of interest-enjoyment? I am mostly marking the question here and acknowledging its difficulty and its risks. I am not comfortable with it, which means that the question boosts my level of interest-enjoyment, even as it leaves me prey to my own work, groping in the labyrinth for a path that is always only emerging.

"And yet . . ."

Genomics and postgenomics have received plenty of hyperenthusiastic adulation and plenty of deflationary critique—both of which, it bears stressing again, are well warranted. But these rather one-dimensional alternatives need to be doubled, at least, read together and against each other, if they are to do justice to the crazy, cool, awesome, exciting, boring, droll, and mad pursuit of postgenomics. As genomics shaded into postgenomics, it became more and more *Toll!*—and that surprised everyone, including its chroniclers—at least this one. Attending to *Toll!*-like expressions of surprise and excitement will enrich our own styles of "dynamic objectivity" in historical and ethnographic analyses of postgenomics, binding our own account to the fuller range of the forces shaping this scientific field.[51] Yes, postgenomicists can be seekers of profit, affirmers of baseless "racial" categories, purveyors of problematic personalized medicines, and all the other personas implicit or explicit in so many ethical, legal, and social studies of this ever-emergent ensemble of scientific practices. But, as a necessary supplement to these necessary accounts, we could stand to come to better terms with the surprise-seeking, creativity-affirming, excitement-purveying dimensions of postgenomicists' personas—even, perhaps especially, if those terms include *Toll!*-like amalgams.

I think there is also a politics to the pursuit of *Toll!*-like affect in contemporary science that is also a necessary supplement to the necessary politics of critique. There is a need in American culture for a greater collective capacity for the surprise-startle and interest-enjoyment that comes through engagement with complex scientific systems, complex arguments, and complex realities. In a time of austerity, compounded by a pervasive devaluation of many kinds of knowledge spiked by elements of outright anti-scientism (in the United States at least), rereading a genealogy of postgenomics could contribute to a cultural need for new idioms for a revalued science that extend beyond its pragmatic applicability, to encompass its ability to provoke widened, shared curiosity about complex biological, cultural, and environmental conditions. My experimental hope is that adding a complex *Toll!*-like amalgam of affects to the diversity of receptors through which we make sense of postgenomics' untimely course of continued emergence will produce, as it has for at least some postgenomicists, surprising effects driven by and embodying of attentive care.

Notes

1. See Keller, "*Drosophila* Embryos as Transitional Objects," and "Developmental Biology as a Feminist Cause?"
2. Rheinberger, "Experimental Systems, Graphematic Spaces."
3. Hansson and Edfeldt, "Toll to Be Paid."
4. Weissmann, "Pattern Recognition and Gestalt Psychology."
5. Quoted in ibid., 2137.
6. Ibid., 2138.
7. Lemaitre notes the element of serendipity involved in these experiments: "I now realize that our success in identifying the function of Toll in the *Drosophila* immune response was partly because we routinely used a mixture of Gram-negative and Gram-positive bacteria to infect flies, whereas other groups only used Gram-negative

bacteria. The Gram-positive bacteria strongly activated the Toll pathway and enabled us to discern the role of Toll"; Lemaitre, "Road to Toll," 524–25.

8. Belvin and Anderson, "Conserved Signaling Pathway."

9. Medzhitov, Preston-Hurlburt, and Janeway, "Human Homologue of the *Drosophila* Toll Protein."

10. Medzhitov, "Approaching the Asymptote."

11. Ibid.

12. Weissmann, "Pattern Recognition and Gestalt Psychology."

13. See Fortun, "Human Genome Project," and *Promising Genomics*.

14. On "epistemic virtues," see Daston and Galison, *Objectivity*, 1st ed. See also M. Fortun, manuscript in preparation.

15. See C. Rice et al., "Developmental Lead Exposure."

16. See Shubin, *Your Inner Fish*.

17. See Tomkins, *Affect, Imagery, Consciousness*; Sedgwick and Frank, "Shame in the Cybernetic Fold"; E. Wilson, "Scientific Interest."

18. Tomkins, *Affect, Imagery, Consciousness*, 273.

19. Ibid.

20. Keller, *Refiguring Life*, 22.

21. Tomkins, *Affect, Imagery, Consciousness*, 187.

22. Ibid., 188.

23. Ibid., 189.

24. Fortun, "Mapping and Making Genes and Histories."

25. Gilbert, "Towards a Paradigm Shift in Biology."

26. It is also a philosopher's question, or at least those philosophers similarly attempting to make some sense of that which has not yet arrived but is in another sense already here: "What Derrida refers to as the 'to come' and Foucault as the 'actual,' Deleuze calls absolute deterritorialization, becoming, or the untimely. It is the pure 'event-ness' that is expressed in every event and, for that reason, immanent in history. It follows that every event raises with greater or lesser urgency the hermeneutic question, 'what happened?'" Patton, "Events, Becoming, and History," 42.

27. Ibid.

28. Ibid.

29. Ibid., emphasis added.

30. Watson, *Avoid Boring People*.

31. Watson, Address at Stated Meeting of the American Academy of Arts and Sciences.

32. Ibid.

33. Author's interview with Robert Moyzis, October 22, 1991. See also Fortun, "Human Genome Project," for my rereading of other comments by Moyzis, regarding his "surprise" at the readiness of many scientists in 1985 to entertain the idea of completely sequencing a human genome.

34. See Fortun, "Human Genome Project."

35. *The Human Genome Project: Hearing Before the Subcommittee on Energy Research and Development, Committee on Energy and Natural Resources, United States*

Senate, 101st Cong. 130 (July 11, 1990) (statement of Peter Domenici, U.S. senator, D-NM).

36. Nelson and White, "Metagenomics and Its Application," 172.

37. Zimmer, "Yet-Another-Genome Syndrome."

38. Ibid.

39. Ibid.

40. Sternberg, "On the Roles of Repetitive DNA Elements," 155.

41. "Bruce Roe on Surprises in the HGP," Oral History Collection, *Cold Spring Harbor Laboratory Digital Archives*, recorded May 29, 2003, http://library.cshl.edu /oralhistory/interview/genome-research/surprises-hgp/roe-surprises-hgp/.

42. "Robert Waterston on Surprises in the HGP," Oral History Collection, *Cold Spring Harbor Laboratory Digital Archives*, recorded June 1, 2003, http://library.cshl .edu/oralhistory/interview/genome-research/surprises-hgp/surprises-hgp/.

43. "Eric Lander on Surprises in the HGP," Oral History Collection, *Cold Spring Harbor Laboratory Digital Archives*, recorded June 2, 2003, http://library.cshl.edu /oralhistory/interview/genome-research/surprises-hgp/lander-surprises-hgp/.

44. "Maynard Olson on Surprises in the HGP," Oral History Collection, *Cold Spring Harbor Laboratory Digital Archives*, recorded June 1, 2003, http://library.cshl .edu/oralhistory/interview/genome-research/surprises-hgp/olson-surprises-hgp/.

45. "Genome, 10 Years Later," A28.

46. Wade, "Decade Later, Genetic Map Yields Few New Cures."

47. P. Edwards, *Vast Machine*.

48. Fujimura and Rajagopalan, "Different Differences."

49. See, e.g., Leonelli and Ankeny, "Re-thinking Organisms."

50. Rheinberger, "Experimental Systems, Graphematic Spaces," 291. *Tatonnement* is the term of François Jacob, who is also the source for Rheinberger's "play of possibilities" (he notes that the English translation of Jacob's book *The Possible and the Actual* does not convey this connotation from the French title, *Le Jeu Possibles*). The Derrida quote is from *Of Grammatology*, 23–24.

51. In this chapter and in my broader work on the sciences, I see myself following up, decades behind, on Evelyn Fox Keller's stunning work that used psychoanalytic theory to show how love was a necessary component of "dynamic objectivity" in the sciences. See Keller, "Dynamic Objectivity."

4

The Polygenomic Organism

John Dupré

This chapter explores some of the ways in which recent insights in genomic science have increasingly problematized the relation between the genome and the phenotype. The focus is less specifically on the human than is that of other chapters in this volume, but the conclusions are of great relevance to the human genome. The complexity of the relation between genotype and phenotype has been explored quite extensively insofar as growing insights into gene/environment interaction, epigenetics, alternative splicing, and so on have made it clear that the extent to which genotype determines phenotype is extremely limited. However, this chapter raises perhaps even deeper concerns about the very idea of the human genome.

The human genome is, of course, generally understood as a set of forty-six chromosomes composed of the three billion or so base pairs in the iconic DNA double helix. The sequencing of these base pairs famously accomplished by the Human Genome Project (HGP) enables us to distinguish this human genome from the genomes of other organisms and also to begin to explore the slight differences between the genomes of individual humans and the ways in which this may account in part for differences between individuals and between particular categories of individuals. The distinctive individual human genome is also supposed to be found in every cell of the individual to whom it pertains, and differences between cells in the individual body are attributed to differences in gene expression, modulated by epigenetic factors, broadly construed. All of this assumes an intimate connection between the individual and the genome that is part of a broader evolutionary perspective in which the phylogenetic relations of individuals and species are completely integrated.

This chapter, however, describes a number of lines of research that are increasingly questioning this intimate relation between genotype and phenotype. Perhaps the most fundamental such development has been the growing understanding of lateral gene transfer, which, especially in the case of the microbial life that arguably dominates the biosphere, has separated the phylogeny of cells from the phylogenies of genomes; even for humans, about half our genome originates from viral sources distinct from the phylogeny standardly represented in the tree of life. This is not the topic of this chapter, however. Here I look rather at the diversity of the genomes found within the contemporary human body and show how contemporary science understands multicellular organisms such as ourselves as containing a multitude of distinct genomes and a diverse collection of what are standardly treated as organisms.

Least radically, I reflect further on the differences between the genomes found in different kinds of human cells—liver cells, nerve cells, skin cells, etc. It is commonly said that these cell types all share the same genome, but the sameness in question is based on a particular and highly abstract characterization of the genome, as no more than a sequence of nucleotides. The actual molecules that constitute the genomes in these cells are very different: the epigenetic changes that affect the different behaviors of these cell types are traceable to the different influences of quite different molecules constituting the genome. It is not always remarked that methylation, the best-known epigenetic alteration of a genome, is actually the conversion of one nucleotide (cytosine) into another (5-methylcytosine). The abstraction of the genome into a four-letter sequence treats these two as the same, but this only emphasizes how abstract is the description on which the judgment of "same genome" is based. Failure to distinguish sufficiently between the abstract representation of the genome as sequence and the far more complex representation in terms of stereochemistry still leads to fundamental misunderstandings of the nature and significance of the genome.[1]

Even in terms of sequence, however, the human body is diverse. Genomes mutate in the course of development, and the mutations are passed to descendant cells. In an unknown proportion of cases human bodies derive from the fusion of two zygotes and form a mosaic of cells with quite distinctive cell types. A pregnant woman is, of course, a genomic mosaic, and fetal cells can remain in the female body long after the end of pregnancy;

in addition, the growing use of transplant medicine, including blood trans-fusion, creates artificial genomic mosaicism. This last factor alone requires that medicine increasingly dispense with the assumption of genomic ho-mogeneity, requiring, most notably, more sophisticated understanding of the immune system.

Finally, and most radically, we are increasingly exploring the signifi-cance of the fact that the human body contains trillions of microbial cells, which in fact make up about 90 percent of the cells in the body. Where once these were thought to be largely opportunistic residents of environ-ments such as the human gut, it is increasingly realized that many or most of these are symbiotic partners, essential to the health and well-being of the composite system. A fascinating aspect of this is that because of the prevalence of gene transfer between microbes, the microbial systems of individuals may vary considerably in what strains of organisms serve these symbiotic functions. Metagenomic analysis, on the other hand, which as-says the total genetic contents of, for example, the gut microbiome, reveals much more homogeneity between individual humans than do traditional ways of classifying the kinds of microbes found therein. What matters is not the strains of microbe that make up the microbiome, but the genetic resources they carry. The full assay of these genetic resources, the Human Microbiome Project, is widely seen as an essential follow-up to, or even completion of, the HGP.

All of these developments show, I argue, that the human is a fundamen-tally polygenomic organism. Even the expression "the human genome" is seen to be problematic, and certainly of much more limited significance than was once thought. And this perspective is crucial to understanding the growing importance for medicine of research in epigenetics, metage-nomics, and other paradigmatically "postgenomic" areas of science.

Genomes and Organisms

Criticisms of the excessive attention on the powers of genes, or "geno-centrism," have been common for many years.[2] While genes, genomes, or more generally DNA are certainly seen as playing a fundamental and even unique role in the functioning of living things, it is increasingly understood that this role can only be properly appreciated when adequate attention is also paid to substances or structures in interaction with which, and only in interaction with which, DNA can exhibit its remarkable powers. Criticisms of genocentrism are sometimes understood as addressing the idea that the genome should be seen as the essence of an organism, the thing or feature that makes that organism what it is. But despite the general decline not only of this idea but of essentialism in general,[3] the assumptions that there is a special relation between an organism and its distinctive genome, and that this is a one-to-one relation, remain largely intact.

The general idea just described might be understood as relating either to types of organisms or to individual organisms. The genome is related to types of organisms by attempts to find within it the essence of a species or other biological kind. This is a natural, if perhaps naïve, interpretation of the idea of the species "bar code," the use of particular bits of DNA sequence to define or identify species membership. But in this paper I am interested rather in the relation sometimes thought to hold between genomes of a certain type and an individual organism. This need not be an explicitly essentialist thesis, merely the simple factual belief that the cells that make up an organism all, as a matter of fact, have in common the inclusion of a genome, and that the genomes in these cells are, barring the odd collision with a cosmic ray or other unusual accident, identical. It might as well be said right away that the organisms motivating this thesis are large multicellular organisms, and perhaps even primarily animals. I shall not be concerned, for instance, with the fungi that form networks of hyphae connecting the roots of plants and are hosts to multiple distinct genomes apparently capable of moving around this network.[4] I should perhaps apologize for this narrow focus. Elsewhere in this chapter I criticize philosophers of biology and others for a myopic focus on a quite unusual type of organism, the multicellular animal.[5] Nonetheless, it is unsurprising that we should have a particular interest in the class of organisms to which we ourselves belong, and this is undoubtedly an interesting kind of organism. And in the end, if my argument is successful for multicellular animals, it will apply all the more easily to other, less familiar forms of life.

At any rate, it is an increasingly familiar idea that we, say, have such a characteristic genome in each cell of our body, and that this genome is something unique and distinctive to each of us. It is even more familiar that there is something, "the human genome," that is common to all of us, although, in light of the first point, it will be clear that this is not exactly the same from one person to another. The first point is perhaps most familiar in the context of forensic genomics, in the realization that the tiniest piece of corporeal material that any of us leaves lying around can be unequivocally traced back to us as its certain source. At any rate, what I aim to demonstrate in this chapter is that this assumption of individual genetic homogeneity is highly misleading and indeed is symptomatic of a cluster of misunderstandings about the nature of the biological systems we denominate as organisms.

Organisms and Clones

A clone, outside *Star Wars*–style science fiction, is a group of cells originating from a particular ancestral cell through a series of cell divisions. The reason we suppose the cells in a human body to share the same genome is that we think of the human body as, in this sense, a clone: it consists of a very large group of cells derived by cell divisions from an originating zygote. A familiar complication is that if I have a monozygotic ("identical") twin, then my twin will be part of the same clone as myself. Although this is only an occasional problem for the human case, in other parts of biology it can be much more significant. Lots of organisms reproduce asexually, and the very expression "asexual reproduction" is close to an oxymoron if we associate biological individuals with clones. For asexual reproduction is basically no more than cell division, and cell division is the growth of a clone. If reproduction is the production of a new individual, it cannot also be the growth of a preexisting individual. Indeed, what justifies taking the formation of a zygote as the initiation of a new organism, reproduction rather than growth, is that it is the beginning of a clone of distinctive cells with a novel genome formed through the well-known mixture between parts of the paternal and the maternal genomes.

As I have noted, it is common to think of genomes as standing in one-to-one relations with organisms. My genome, for instance, is almost surely unique, and it, or something very close to it, can be found in every cell in my body. Or so, anyhow, the standard story goes. The existence of clones that do not conform to the simple standard story provides an immediate and familiar complication for the uniqueness part of this relation. If I had a monozygotic ("identical") twin, then there would be two organisms whose cells contained (almost) exactly the same genomes; we would both, since originating from the same lineage-founding zygote, be parts of the same clone. And lots of organisms reproduce asexually all or some of the time, so this difficulty is far from esoteric.

Some biologists, especially botanists, have bitten the bullet here. They distinguish ramets and genets, where the genet is the sum total of all the organisms in a clone, whereas the ramet is the more familiar individual.[6] Thus, a grove of trees propagated by root suckers, such as are commonly formed, for instance, by the quaking aspen (*Populus tremuloides*), in the deserts of the Southwest United States, is one genet but a large number of ramets. Along similar lines, it has famously been suggested that among the largest organisms are fungi of the genus *Armillaria*, the familiar honey fungi.[7] A famous example is an individual of the species *Armillaria ostoyae* in the Malheur National Forest in Oregon that was found to cover 8.9 km² (2,200 acres).[8] To the average mushroom collector, a single mushroom is an organism, and it would be strange indeed to claim that two mushrooms collected miles apart were parts of the same organism. There is nothing wrong with the idea that for important theoretical purposes this counterintuitive conception may be the right one; there is also nothing wrong with the more familiar concept of a mushroom. The simple but important moral is just that we should be pluralistic about how we divide the biological world into individuals: different purposes may dictate different ways of carving things up.

It is pretty clear, however, that we cannot generally admit that parts of a clone are parts of the same individual. Whether or not there are technical contexts for which it is appropriate, I doubt whether there are many interesting purposes for which two monozygotic human twins should be counted as two halves of one organism. Or anyhow, there are certainly interesting purposes for which they must be counted as distinct organisms, including almost all the regular interests of human life. An obvious reason for this is that most of the career of my monozygotic twin (if I had one) would be quite distinct from my own. And for reasons some of which should become clearer in light of the discussion below of epigenetics, the characteristics of monozygotic twins tend to diverge increasingly as time passes. The careers of monozygotic twins may carry on independently from birth in complete ignorance of one another, but it is hardly plausible that if I were now to discover that I had a monozygotic twin, this would drastically change my sense of who I was (i.e., a spatially discontinuous rather than spatially connected entity). Some kind of continuing connection seems needed even to make sense of the idea that these could be parts of the same thing. Being parts of the same clone is at any rate not a sufficient condition for being parts of the same biological individual.

However, we should not immediately assume that the concept of a genet encompassing a large number of ramets is generally indefensible. A better conclusion to draw is that theoretical considerations are insufficient to determine unequivocally the boundaries of biological objects. Sometimes, perhaps always, this must be done relative to a purpose. There are many purposes for which we distinguish human individuals and for the great majority of which it would make no sense to consider my twin and myself part of the same entity. My twin will not be liable to pay my debts or care for my children, for instance, if I should default on these responsibilities, though it is interesting in the latter case that standard techniques for determining that they are my children would not distinguish my paternity from my twin's. This may even point to an evolutionary perspective from which we are best treated as a single individual. And when it comes to the trees, this is surely the right way to go. For the purposes of some kinds of evolutionary theory the single genet may be the right individual to distinguish, but if one is interested in woodland ecology, what matters will be the number of ramets. If this seems an implausible move, this is presumably because of the seemingly self-evident individuality of many biological entities. I hope that some of the considerations that follow will help to make this individuality a lot less self-evident than it might appear at first sight. But whether or not the pluralism I have suggested for individual boundaries is defensible, the assumption of a one-to-one relation between genomes and organisms is not. I will explain the objections to this assumption in what I take to be an order of increasing fundamentality. At any rate, as the next section shows, the various phenomena of genetic mosaicism suffice to demonstrate that genotypes will not serve to demarcate the boundaries of biological individuals. In other words, genomic identity is not a necessary condition for being part of the same biological individual.

Genomic Chimeras and Mosaics

The general rubric of genomic mosaicism encompasses a cluster of phenomena. An extreme example, sometimes distinguished more technically as chimerism, is of organisms that have resulted from the fusion of two zygotes, or fertilized eggs, in utero. The consequence of this is that different parts of the organism will have different genomes—the organism is a mosaic of cells of the two different genomic types from which it originated. A tragic consequence of this has been the occasional cases of women who have been denied custody of their children on the basis of genetic tests that appeared to show that they and the child were not related. It has turned out that the explanation of this apparent contradiction of a connection of which the mother was in no doubt was that she was a genomic mosaic of this kind, and the cells tested to establish the parental relation were from a different origin than the gametes that gave rise to the child.[9] With the exception of a modest degree of chimerism found in some fraternal twins who have exchanged blood and blood cell precursors in utero and continue to have distinct genotypes in adult blood cells, such cases are generally assumed to be very rare in humans. However, chimeras do not necessarily experience any unusual symptoms, so the prevalence of full chimerism, chimerism derived from multiple zygotes, is not really known and may be much higher than suspected.

Probably more common than chimerisms resulting from the fusion of two zygotes are those resulting from mutations at some early stage of cell division. One well-known example of this is XY Turner syndrome, in which the individual is a mixture of cells with the normal XY karyotype, the complement of sex chromosomes found in most males, and XO cells, in which there is no Y chromosome and only one X chromosome.[10] Turner syndrome is a condition of girls in which all the cells are XO (i.e., with one X chromosome missing, as opposed to the standard XX); people with XY Turner syndrome generally have normal male phenotypes, though a small percentage are female and a similar small percentage are intersexed. The large majority of fetuses with either condition are spontaneously aborted. The phenotype displayed by XY Turner cases is presumably dependent on exactly when in development the loss of the X chromosome occurs.

Chimerism is quite common in some other organisms. When cows have twins, there is usually some degree of shared fetal circulation, and both twins become partially chimeric. This has been familiar from antiquity in the phenomenon of freemartinism, freemartins being the sterile female cattle that have been known since the eighteenth century invariably to have a male twin. This is the normal outcome for mixed-sex bovine twins, with the female twin being masculinized by hormones deriving from the male twin.[11] This has occasionally been observed in other domesticated animals. Even more so than for the human case, the prevalence of this and other forms of chimerism in nature is not known.

The chimeras mentioned so far are all naturally occurring phenomena. Much more attention has lately been paid to the possibility of artificially producing chimeras in the laboratory. And unsurprisingly, the most attention has been focused on the possibility of producing chimeras, or hybrids, that are in part human. Recent controversy has focused on the ethical acceptability of generating hybrid embryos for research purposes by transplanting a human nucleus into the egg cell of an animal of another species, usually a cow.[12] Since all the nuclear DNA in such a hybrid is human, it can be argued that this is not a chimera at all, at least in the genetic sense under consideration. On the other hand, such cells will contain nonhuman DNA in the mitochondria, the extranuclear structures in the cell that provide the energy for cellular processes.[13] No doubt the mixture of living material from humans and nonhumans is disturbing to many whether or not the material in question is genetic, as is clear from controversy over the possibility of xenotransplantation, use of other animals to provide replacement organs. But this is not my concern in the present chapter.[14]

Modern laboratories, at any rate, are well able to produce chimeric organisms. At the more exotic end of such products, and certainly chimeras, are such things as "geeps," produced by fusing a sheep embryo with a goat embryo. The adults that develop from these fused embryos are visibly and bizarrely chimeric, having sheep wool on parts of their bodies and goat hair on others. Much more significant, however, are the transgenic organisms that have caused widespread public discomfort in the context of genetically modified (GM) foods.[15] These are often seen as some kind of violation of the natural order, the mixing together of things that nature or God intended to keep apart.[16] Whatever other objections there may be to the production of GM organisms, it will become increasingly clear that this is not one with which I am sympathetic: organisms do not naturally display the genetic purity that this concern seems to assume.

The chimeric organisms discussed so far in this section have been organisms originating to some degree from two distinct zygotes. (The exception is the XY Turner syndrome, which should strictly have been considered in the context of the following discussion.) Other cases relevant to the general topic of intraorganismic genomic diversity, but generally referred to by the term "mosaicism" rather than "chimerism," exhibit genomic diversity but deriving from a single zygotic origin. Such mosaicism is undoubtedly very common. One extremely widespread instance is the mosaicism common to most or all female mammals that results from the expression of different X chromosomes in different somatic cells. In the human female, one of the two X chromosomes in each cell is condensed into a cellular object referred to as the Barr body and is largely inert. Different parts of the body may have different X chromosomes inactive, implying that they have different active genotypes. This phenomenon will apply to most sexually reproducing organisms, though in some groups of organisms (e.g., birds) it is the male rather than the female that is liable to exhibit this kind of mosaicism.[17] The most familiar phenotypic consequence of this phenomenon is that exhibited by tortoiseshell or calico cats, in which the different coat colors reflect the inactivation of different X chromosomes. Although there are very rare cases of male calico cats, these appear to be the result of chromosomal anomaly (XXY karyotype), chimerism, or mosaicism in which the XXY karyotype appears as a mutation during development.[18]

Returning to chimerism, mosaicism deriving from distinct zygotes, a quite different but very widespread variety is exhibited by females, including women, after they have borne children and is the result of a small degree of genomic intermixing of the maternal and offspring genomes. Though scientists have been aware of this phenomenon for several decades, it has recently been the focus of increased attention for several reasons. For example, recent work suggests that the transfer of maternal cells to the fetus may be important in training the latter's immune system.[19] Another reason for increasing interest in this topic is the fact that it opens up the possibility of genetic testing of the fetus using only maternal blood, thus avoiding the risks inherent in invasive techniques for fetal testing such as amniocentesis.[20] It should also be noted that maternal cells appear to persist in the offspring and vice versa long after birth, suggesting that we are all to some degree genomic mosaics incorporating elements from our mothers and, for women, our offspring.

One final cause of chimerism that must be mentioned is the artificial kind created by transplant medicine, including blood transfusions. Very likely this will continue to become more common as techniques of transplantation become more refined and successful. A possibility increasingly under discussion is that this will eventually be extended, through the development of xenotransplantation, to include interspecific mosaicism. At any rate, any kind of transplantation, except that involving cells produced by the recipient himself or herself, will produce some genomic chimerism. So, in summary, both natural and artificial processes, but most commonly the former, generate significant degrees of chimerism in many, perhaps almost all, multicellular organisms, including ourselves. The assumption that all the cells in a multicellular organism share the same genome is therefore seriously simplistic, and, as mentioned above, conclusions drawn from this simplistic assumption, for example, about the violation of nature involved in producing artificial chimeras, are, to the extent that they rely on this assumption, ungrounded.

Epigenetics

The topics of chimerism and mosaicism so far discussed address the extent to which the cells that make up a body are genomically uniform in the sense of containing the same DNA sequences. This discussion runs a risk of seeming to take for granted the widely held view that, given a certain common genome, understood as a genome with a particular sequence of nucleotides (the As, Cs, Gs, and Ts familiar to everyone in representations of DNA sequence), the behavior of other levels of biological organization will be determined. Perhaps a more fundamental objection to the one genome–one organism doctrine is that this common assumption is entirely misguided. The reason that the previous discussion may reinforce such an erroneous notion is that the comparisons and contrasts between genomes were implicitly assumed to be based entirely on sequence comparisons. But to know what influence a genome will actually have in a particular cellular context, one requires a much more detailed and nuanced description of the genome than can be given merely by sequence. And once we move to that more sophisticated level of description, it becomes clear that, even within the sequence-homogeneous cell lineages often thought to constitute a multicellular organism, there is a great deal of genomic diversity. These more sophisticated descriptions are sought within the burgeoning scientific field of epigenetics, or epigenomics.

A good way of approaching the subject matter of epigenetics is to reflect on the question why, if indeed all our cells do have the same genome, they nevertheless do a variety of very different things. It is of course very familiar that not all the cells in a complex organism do the same things—they are differentiated into skin cells, liver cells, nerve cells, and so on. Part of the explanation for this is that the genome itself is modified during development, a process studied under the rubric of epigenetics or epigenomics.[21] The best-known such modification is methylation, in which a cytosine molecule in the DNA sequence is converted to 5-methylcytosine, a small chemical addition to one of the nucleotides, or bases, that make up the DNA molecule. This has the effect of blocking transcription of the DNA sequence at particular sites in the genome. Other epigenetic modifications affect the protein core, or histones, which form part of the structure of the chromosome, and also influence whether particular lengths of DNA are transcribed into RNA. It is sometimes supposed that these are not "real," or anyhow significant, alterations of the genome, perhaps because we still describe the genome sequence in the same way, referring to either cytosine or 5-methylcytosine by the letter C. But all this really shows is that the standard four-letter representation of genomic sequence is an abstraction. As a matter of fact, there are about twenty nucleotides that can occur in DNA sequences, and it is only our choice of representation that maintains the illusion that some chemically fixed entity, the genome, can be found in all our cells. If we were to change the representation to a more fine-grained description of chemical composition, we would find a much greater genomic diversity than is disclosed by the more abstract and familiar four-letter code.

It is true that part of the value of the abstraction that treats the genome as consisting of only four nucleotides is that this does represent a very stable underlying structure. This has provided extremely useful applications that use stable genome sequence to compare or identify organisms, applications ranging from phylogenetic analysis to forensic DNA fingerprinting. Phylogenetic analysis, the investigation of evolutionary relations between kinds of organisms, here depends on the stability of genomes as they are transmitted down the generations, and DNA fingerprinting depends on the admittedly much shorter term stability of genome sequence within the life of the individual. Methylation, on the other hand, is reversible and often reversed. However, overemphasis on this stable core can be one of the most fundamental sources of misunderstanding in theoretical biology.

Such misunderstanding is sometimes expressed in the so-called central dogma of molecular biology.[22] This is generally interpreted as stating that information flows from DNA to RNA to proteins, but never in the reverse direction. I do not wish to get involved in exegesis of what important truth may be alluded to with this slogan, and still less into the vexed interpretation of the biological meaning of "information."[23] What is no longer disputable is that causal interaction goes both in the preferred direction of the central dogma and in the reverse direction. Epigenetic changes to the genome are induced by chemical interactions with the surrounding cell (typically with RNA and protein molecules). A reason why this is so important is that it points to a mechanism whereby even very distant events can eventually have an impact on the genome and its functioning. The classic demonstration of this is the work of Michael Meaney and colleagues, on ways in which maternal care can modify the development of cognitive abilities in baby rats, something that has been shown to be mediated by methylation of genomes in brain cells.[24] The most recent work by this group has provided compelling reasons to extrapolate these results to humans.[25] Whether epigenetic research shows that genomes are diverse throughout the animal body of course depends on one's definition of "genome" and one's criterion for counting two as the same. It needs just to be noted that if we choose a definition that, *pace* the points made in earlier sections, counts every cell as having the same genome, we will be overlooking differences that make a great difference to what the cell actually does.

Symbiosis and Metaorganisms

In this section I want to make a more radical suggestion. So far I have considered the diversity of human (or other animal) cells that may be found in an individual organism, and the phenomena I have described are generally familiar ones to molecular biologists. In this section I propose that there are good reasons to deny the almost universal assumption that all the cells in an individual must belong to the same species. This may seem no more than tautological: if a species is a kind of organism, then how can an organism incorporate parts or members of different species? The resolution of this paradox is to realize that very general terms in biology such as "species" or "organism" do not have univocal definitions: in different contexts these terms can be used in different ways. For the case of species, this is quite widely agreed among philosophers of biology today.[26] I am also inclined to argue something similar for organisms. Very roughly, I want to suggest that the organisms that are parts of evolutionary lineages are not the same things as the organisms that interact functionally with their biological and nonbiological surroundings. The latter, which I take to be more fundamental, are composed of a variety of the former, which are the more traditionally conceived organisms. But before explaining this idea in more detail, I need to say a bit more about the facts on which it is based. I introduce these with specific reference first to the human.

A functioning human organism is a symbiotic system containing a multitude of microbial cells—bacteria, archaea, and fungi—without which the whole would be seriously dysfunctional and ultimately nonviable. Most of these reside in the gut, but they are also found on the skin and in all body cavities. In fact, about 90 percent of the cells that make up the human body belong to such microbial symbionts, and, owing to their great diversity, they contribute something like 99 percent of the genes in the human body. It was once common to think of these as little more than parasites, or at best opportunistic residents of the various vacant niches provided by the surfaces and cavities of the body. However, it has become clear that, on the contrary, these symbionts are essential for the proper functioning of the human body. This has been recognized in a major project being led by the U.S. National Institutes of Health (NIH) that aims to map the whole set of genes in a human, the Human Microbiome Project.[27]

The role of microbes in digestion is most familiar and is now even exploited by advertisers of yogurt. But even more interesting are their roles in development and in the immune system. In organisms in which it is possible to do the relevant experiments, it has turned out that genes are activated in human cells by symbiotic microbes, and vice versa.[28] Hence, the genomes of the human cells and the symbiotic microbes are mutually dependent. And it seems plausible that the complex microbial communities that line the surfaces of the human organism are the first lines of defense in keeping out unwanted microbes.[29] Since the immune system is often defined as the system that distinguishes self from nonself, this role makes it particularly difficult to characterize our symbiotic partners as entirely distinct from ourselves. Finally, it is worth recalling that we are not much tempted to think of the mitochondria that provide the basic power supply for all our cellular processes as distinct from ourselves, yet these are generally agreed to be long-captive bacteria that have lost the ability to survive outside the cell.

These phenomena are far from being unique to the human case, and arguably similar symbiotic arrangements apply to all multicellular animals. In the case of plants, the mediation of the metabolic relations between the plant roots and the surrounding soil is accomplished by extremely complex microbial systems involving consortia of bacteria, as well as fungi whose webs pass in and out of the roots, and which are suspected of transferring nutrients between diverse members of the plant community, suggesting a much larger symbiotic system.[30]

My colleague Maureen O'Malley and I have suggested that the most fundamental way to think of living things is as the intersection of lineages and metabolism.[31] The point we are making is that, contrary to the idea that is fundamental to the one genome–one organism idea, the biological entities that form reproducing and evolving lineages are not the same as the entities that function as wholes in wider biological contexts. Functional biological wholes, the entities that we primarily think of as organisms, are in fact cooperating assemblies of a wide variety of lineage-forming entities. In the human case, as well as what we more traditionally think of as human cell lineages, these wider wholes include a great variety of external and internal symbionts. An interesting corollary of this perspective is that although we do not wish to downplay the importance of competition in the evolution of complex systems, the role of cooperation in forming the competing wholes has been greatly underestimated. And there is a clear tendency in evolutionary history for entities that once competed to form larger aggregates that now cooperate.

Conclusion

It should be clear that there is a continuity between the phenomena I described under the heading of chimerism and mosaicism and those discussed in the preceding section. Living systems, I am arguing, are extremely diverse and opportunistic compilations of elements from many distinct sources. These include components drawn from what are normally considered members of the same species, as illustrated by many of the cases of chimerism, but also, and more fundamentally, by the collaborations between organisms of quite different species, or lineages, which have been the topic of the preceding section. All of these cases contradict the common if seldom articulated assumption of one genome, one organism.

One plausible hypothesis about the attraction of the one genome–one organism assumption is that it represents an answer to the question, what is the *right* way of dividing biological reality into organisms? But, as I have argued throughout this chapter, there is no unequivocal answer to this question. From the complex collaborations between the diverse elements in a cell, themselves forming in some cases (such as mitochondria) distinct lineages, through the intricate collaborations in multispecies microbial communities, to the even more complex cooperations that compose multicellular organisms, biological entities consist of disparate elements working together. Different questions about, or interests in, this hierarchy of cooperative and competitive processes will require different distinctions and different boundaries defining individual entities. As with the more familiar question about species, in which it is quite widely agreed that different criteria of division will be needed to address different questions, so it is, I have argued, with individuals. This is one of the more surprising conclusions that have emerged from the revolution in biological understanding that is gestured at by the rubric, "genomics."

Returning finally to the distinctively human, the capacities that most clearly demarcate humans from other organisms—language, culture—are the capacities that derive from our increasing participation in ever more complex social wholes. A further extension of the argument sketched in the preceding paragraph would see this as the next stage in the hierarchy of collaboration and perhaps, as has often been speculated, genuinely marking the human as a novel evolutionary innovation. Rather less speculatively, it is arguably a striking irony that the often-remarked centrality of individualism in the past two hundred years of social theory has perhaps been the greatest obstacle to seeing the profoundly social, or anyhow cooperative, nature of life more generally.

Notes

"The Polygenomic Organism," by John Dupré, first appeared in *The Sociological Review* 58 (2010) and is printed by permission of the publisher, John Wiley and Sons. © 2010 The Author. Editorial organization © 2010 The Editorial Board of *The Sociological Review*.

1. Barnes and Dupré, *Genomes and What to Make of Them*.
2. For a recent example, see ibid.

3. This applies, at any rate, among philosophers concerned with the details of scientific belief. Essentialism has had something of a resurgence among more abstractly inclined metaphysicians; see Ellis, *Scientific Essentialism*; and Devitt, "Resurrecting Biological Essentialism."

4. Sanders, "Ecology and Evolution of Multigenomic Arbuscular Mycorrhizal Fungi."

5. O'Malley and Dupré, "Size Doesn't Matter."

6. Harper, *Population Biology of Plants*.

7. Smith, Bruhn, and Anderson, "Fungus *Armillaria bulbosa*."

8. See Casselman, "Strange but True."

9. Yu et al., "Disputed Maternity Leading to Identification of Tetragametic Chimerism."

10. O. Edwards, "Masculinized Turner's Syndrome XY-XO Mosaicism."

11. Exactly to what extent this is the normal outcome remains, as with so many phenomena in this area, somewhat unclear, however; see Zhang et al., "Diagnosis of Freemartinism in Cattle."

12. Although research involving hybrid embryos is generally thought unacceptable unless there are clear potential medical benefits, opinion in the United Kingdom is quite finely divided on this topic; see Jones, "What Does the British Public Think about Human-Animal Hybrid Embryos?"

13. As a matter of fact, the mitochondria are now known to be descendants of bacteria that long ago became symbiotically linked to the cells of all eukaryotes, or "higher" organisms. This may suggest a further sense in which we are all chimeric, a suggestion I elaborate shortly.

14. But see Parry, "Interspecies Entities and the Politics of Nature"; and Twine, "Genomic Natures Read through Posthumanisms."

15. Milne, "Drawing Bright Lines."

16. Barnes and Dupré, *Genomes and What to Make of Them*.

17. Curiously, however, it appears that birds find less need to compensate for the overexpression of genes on the chromosome of which one sex has two (in birds the male has two Z chromosomes). So this kind of mosaicism will be less common, or may not occur at all; see Marshall Graves and Disteche, "Does Gene Dosage Really Matter?"

18. Centerwall and Benirschke, "Male Tortoiseshell and Calico (T—C) Cats."

19. Mold et al., "Maternal Alloantigens Promote the Development."

20. Lo, "Fetal DNA in Maternal Plasma"; Benn and Chapman, "Practical and Ethical Considerations of Noninvasive Prenatal Diagnosis."

21. It appears that the phenomenon in question may not be fully explicable at all, however, as gene expression is also importantly affected by random processes, or noise; see Raser and O'Shea, "Noise in Gene Expression." But there is also growing evidence that noise of this kind may be adaptive, and hence this effect may have been subject to natural selection; see Maamar, Raj, and Dubnau, "Noise in Gene Expression."

22. This phrase was introduced originally by Francis Crick, and I have no wish

to accuse Crick himself of misunderstanding. Indeed, the use of the word "dogma" suggests a degree of irony.

23. Maynard Smith, "Concept of Information in Biology"; Griffiths, "Genetic Information."

24. Champagne and Meaney, "Stress during Gestation."

25. McGowan et al., "Epigenetic Regulation of the Glucocorticoid Receptor."

26. For discussion see various essays in R. Wilson, *Species*.

27. See the NIH website for the Human Microbiome Project, available as of January 7, 2014, at http://nihroadmap.nih.gov/hmp/.

28. Rawls, Samuel, and Gordon, "Gnotobiotic Zebrafish."

29. More traditional views of the limits of the human organism might make it seem strange that a strong correlate of infection with the hospital superbug *Clostridium difficile* is exposure to powerful courses of antibiotics, though this correlation is not quite as pervasive as was earlier thought; see Dial et al., "Patterns of Antibiotic Use."

30. Hart, Reader, and Klironomos, "Plant Coexistence Mediated by Arbuscular Mycorrhizal Fungi."

31. Dupré and O'Malley, "Varieties of Living Things."

5

Machine Learning and Genomic Dimensionality

FROM FEATURES TO LANDSCAPES

Adrian Mackenzie

The Google Compute Engine, a globally distributed ensemble of computers, was briefly turned over to exploration of cancer genomics during 2012
and publicly demonstrated during the annual Google I/O conference. Midway through the demonstration, in which a human genome is visualized
as a ring in "Circos" form (see chap. 6 of this volume),[1] the speaker, Urs
Hölzle, senior vice president of infrastructure at Google, "then went even
further and scaled the application to run on 600,000 cores across Google's
global data centers."[2] The audience clapped. The world's "3rd largest super-
computer," as it was called by TechCrunch, a prominent technology blog,
"learns associations between genomic features."[3] We are in the midst of
many such demonstrations of "scaling applications" of data in the pursuit
of associations between "features."

Nearly all contemporary epistemic practices—scientific or not—depend
on infrastructures of communication and computation. But some sciences
are much more extensively invested in computing infrastructures as ways
of making sense of data. These investments are not incidental, ancillary,
or superficial. They deeply permeate the practices, the imaginaries, and
the modes of existence of high-profile, large-scale sciences. As the Google
Compute demonstration suggests, genomes feature strongly in the data
economy. The abundance, size, complexity, and subtle variability of genomes
as *sequence data* are a touchstone for the more general economization of data
we are currently experiencing. Genomic infrastructures—the ensemble of

software, hardware, algorithms, networks, and repositories that handle sequence data and other biological data—tell us something of how genomes come to be what they are. The power of computational machines such as the Google Compute Engine to learn associations in genomes is widely held to be the only way to make sense of the vast aggregates of sequence data produced by genomics, especially since the advent of high-throughput sequencing platforms in the mid-2000s. In classifying, predicting, and exploring biological processes through sequence data, *machine learning* makes sense of differences and variations that cannot be seen or grasped by eye or hand alone.

Such techniques—and I describe a few examples below—also change what a genome is for us. The mode of existence of genomes, ranging across epistemic (see chap. 2 of this volume), affective (see chap. 3 of this volume), political (see chap. 11 of this volume), and ontological (see chap. 4 of this volume) aspects, changes as these techniques come into play. A complicated information ecology has developed in and around genomics over the past few decades. I think we need to understand better how the heavy investment in information retrieval systems for biological data has gradually produced a substrate of biological data that supports shifts in epistemic and economic dynamics of genomes. The spectacle of the Google Compute demonstration attests to some of those shifts. Machine learning techniques crystallize this transformation in their heavy dependence on information retrieval systems and their almost aggressive efforts to reorganize scientific knowing on a different scale. Machine learning does not exhaust or indeed come anywhere near replacing the sophisticated understandings of biological processes associated with epigenetics or with pervasive transcription (see chap. 2 of this volume). But it cuts a swathe through scientific literatures, through the training and recruitment of scientists, and through public and private investments in genomic science, and it connects genomics to broader social, technological, economic, and political dynamics.

Looking outward, contemporary genomics exemplifies and possibly points to some of the limits of "data-intensive" science, government, media, and business more generally. The attempts to machine-learn genomes illustrate how difficult it is to fully domesticate data or render them fully subject to calculation. In this chapter I use code and data vignettes from contemporary genomics to explore how genomic data practice increasingly resorts to Google Compute–style modes of operation. Writing code and working with genomic data as a social scientist or humanities scholar are ways to explore and perhaps think through some of the transformations associated with data practice more generally. At times, the code and data will be hard to read. I suggest that readers might treat those moments of illegibility as encounters with the often-convoluted materialities of genomics.

"Wide, Dirty, and Mixed" Data Sets in Genomics

The I/O conference audience, largely comprising software developers, could hardly be expected to have a detailed interest in what was being shown on the screen. Their interest was steered toward the immediate availability of huge computing power: from 10,000 to 600,000 cores in a few seconds. The principal chain of associations in the genomics demonstration was, presumably, something like genome → complexity → cancer/disease → life/death → important. Yet the Google Compute demonstration is, I would suggest, typical of some recent transformations in how genomes are handled and in what is being addressed in efforts to analyze genomic data. This transformation is only hinted at in the Google I/O keynote address in Hölzle's talk of genomic features, gene expression, and patient attributes. The only concrete indication of what was happening in the demonstration consisted in the mention of one machine learning algorithm, RF-ACE (Random Forest–Artificial Contrasts with Ensembles). As Google's web pages described it,

> It allows researchers to visually explore associations between factors such as gene expression, patient attributes, and mutations—a tool that will ultimately help find better ways to cure cancer. The primary computation that Google Compute Engine cluster performs is the RF-ACE code, a sophisticated machine learning algorithm which learns associations between genomic features, using an input matrix provided by ISB (Institute for Systems Biology). When running on the 10,000 cores on Google Compute Engine, a single set of associations can be computed in seconds rather than ten minutes, the time it takes when running on ISB's own cluster of nearly 1000 cores. The entire computation can be completed in an hour, as opposed to 15 hours.[4]

Amid this mire of fairly technical computing jargon, we might observe that Google applies here an algorithm developed by engineers at Intel Corporation and Amazon to genomic data sets provided by the Institute of Systems Biology (ISB), Seattle, a doyen of big-data genomics. This confluence of commerce (Amazon), industry (Intel), media (Google), and genomic science (ISB) is, I suggest, symptomatic of data practices in genomics and elsewhere.

Now the RF-ACE algorithm, first published in 2009, is a state-of-the-art attempt to deal with "modern data sets" that are "wide, dirty, mixed with both numerical and categorical predictors, and may contain interactive effects that require complex models."[5] Such algorithms and the data sets they work on have a distinctive texture, which we should try to grasp if we want to see what is happening to genomes. Coming to grips with the ways in which algorithms traverse, shape, and partition or segment the notoriously monotonous and superabundant sequence data that genomics has been generating for the past few decades might be a useful way to make sense of recent developments in genomics more generally.

We can see what is happening inside RF-ACE because Google has recently made the RF-ACE algorithm accessible as a code library for the widely used statistical programming language R.[6] In making sense of the reorganization of sciences within a knowledge economy, being able to access and follow the transformations genome-related data undergo in contemporary algorithms is a definite good. Unlike some other places where machine learning is used (national intelligence, for instance), we can follow what happens to the data. With a little knowledge of programming languages such as R or Python used by genomic scientists, it is possible to download the package (https://code. google.com/p/rf-ace/wiki/RPackageInstallation), experiment with the algorithm, and thereby gain some sense of the texture of contemporary data practices in fields such as genomics, but also business and others.

The beginning of an R session with RF-ACE, using a sample data set whose shape and composition are typical of genomic data sets, is shown below:

```
N:output C:class N:input N:noise_1 N:noise_2 N:noise_3 N:noise_4
sample_1 1.586 3 7.490 −0.220 −0.078 0.078 −0.166 sample_2 3.971
1 5.832 −0.357 0.097 0.684 −2.374 sample_3 6.202 NaN 6.477 −1.420
0.430 0.474 0.450
N:noise_5 C:noise_6 N:noise_7
sample_1 1.464 1 −0.882
sample_2 1.032 3 −0.104
sample_3 1.333 3 −0.055
Predictions. . . .
[[1]]
[[1]]$target
[1] "N:output"
[[1]]$sample
[1] "sample_1"
```

```
[[1]]$true
[1]  1.586
[[1]]$prediction
[1]  1.635
[[1]]$error
[1]  1.223
```

I am not going to delve into the RF-ACE algorithm here, but already in this output from the code, several typical machine learning features appear. First, the algorithm produces "predictions" whose values can be compared to the actual values of the data. Prediction plays a crucial role in working with genomic data, not only in the sense that one might wish to predict who will develop Hodgkin's lymphoma on the basis of a DNA sequence produced from a saliva swab, but also in the sense that one might want or need to predict how a particular domain or region of a genome functionally interacts with various biological processes.

Second, we can see here something of the shape and composition of contemporary data sets—they contain numerical data (N:class), classifications (C:class), and missing values ("NaN"—not a number). What these different fields of data refer to is *almost* completely irrelevant. Field 3 (Class:N) might refer to population group, type of cancer, species, and so on. Field 4 might be a cholesterol level, concentration of an enzyme, average survival time, and so on. Even at a generic level, this data extract looks different from the typical data sets of modern statistics. Compare, for instance, the "iris" data set, one of the most heavily used data sets in all of statistics, dating from the late 1930s, and extensively analyzed in the context of plant breeding and genetics by R. A. Fisher, a key contributor to twentieth-century statistics (see table 5.1).[7]

While Fisher developed various ways of discriminating between different species of iris in terms of petal and sepal lengths, we can already see from the extract of the data set that it is shaped differently than the RF-ACE data set. It is quite narrow as it has only a few columns, the data are nearly all of one type (lengths), and the data are clean (there are no missing values). If the RF-ACE data set is typical of contemporary genomic data, iris, in its relative homogeneity, is typical of classical statistics, and much genetic data prior to genomics. Third and finally, there are many elements missing from the RF-ACE data because they cannot be shown or easily displayed. In this example, the actual data set is *wider* than it is long (that is, 103 samples, each characterized by three hundred different "features" or variables), and this breadth of data simply cannot be displayed on the page. *Wide* data sets

Table 5.1. Iris Data Set

	Sepal.Length	Sepal.Width	Petal.Length	Petal.Width	Species
1	5.1	3.5	1.4	0.2	setosa
2	4.9	3.0	1.4	0.2	setosa
3	4.7	3.2	1.3	0.2	setosa
4	4.6	3.1	1.5	0.2	setosa
5	5.0	3.6	1.4	0.2	setosa

are quite common in machine learning settings generally, but particularly common in genomics, where there might only be a relatively small number of biological samples but a huge amount of sequence data for each sample. We will soon see how machine learning seeks to corral and contain their tendencies to sprawl. In these features—predictivity, heterogeneity, expansiveness—we glimpse some problems that are not limited to, but certainly writ particularly large in, genomics.

In order to see how machine learning comes to bear on genomes, we need to see how such wide and mixed data sets have become available. The iconic data type in genomics is the sequence of DNA base pairs, often known as the "base sequence." The abundance of base sequence data is well known and has been an integral, material component of genomes and of genomics. Even if sequencing was and is boring (see chap. 3 of this volume), generating sequence data has been and remains the genomic modus operandi.

```
ctttggatccctactaggcctatgcttaattactcaaatcatcacaggcct
cctactagctatgcactacacagccgacactagcctggccttctcct
ccgtcgcccacatatgccgggacgtacaattcggctgactcatccgtaacct
```

In a sense, the sequence shown above, the first 150 bases from the GenBank sequences for gene U15717 from the tanager (*Ramphocelus*), a bird that lives at the edge of tropical forests, does not fit the picture I have just been drawing. These data are monotonously flat and unvarying, it seems. It might be wide in the sense described above, but it is not mixed. Importantly, however, the abundance of sequence data has attracted and facilitated the many tools and databases of bioinformatics, the hybrid discipline that combines computer science and genomic biology. A teeming ecology of databases and software tools focuses on adding layers and annotations to such sequences. This means that base sequences not only accumulate but also are transformed in this accumulation. What we might, following

Marx, call the "primitive accumulation" phase of genomics over the past 25 years has yielded not only a highly accessible stock of sequence data but sequences that can be mapped onto, annotated, tagged, and generally augmented by many other forms of data. Again, the software tools of bioinformatics implement the combination of retrievability and augmentation.

```
library(biomart)
marts <- listMarts()
print(head(marts))
biomart version
1 ensembl ENSEMBL GENES 69 (SANGER UK)
2 snp ENSEMBL VARIATION 69 (SANGER UK)
3 functional_genomics ENSEMBL REGULATION 69 (SANGER UK)
4 vega VEGA 49 (SANGER UK)
5 bacteria_mart_16 ENSEMBL BACTERIA 16 (EBI UK)
6 fungi_mart_16 ENSEMBL FUNGI 16 (EBI UK)
```

The code fragment above loads the evocatively named "biomart" library and lists the first ten available "biomarts." There are sixty-six biomarts online at the time of writing. Already these names—"ensembl," "snp," "vega"—suggest that these databases cover a wide range of biological data, some of which are sequence based. Looking more closely at one of the more sequence-oriented databases:

```
[1]  "oanatinus_gene_ensembl"  "tguttata_gene_ensembl"
[3] "cporcellus_gene_ensembl"  "gaculeatus_gene_ensembl"
[5] "lafricana_gene_ensembl"  "itridecemlineatus_gene_ensembl"
[7] "mlucifugus_gene_ensembl"  "hsapiens_gene_ensembl"
[9] "choffmanni_gene_ensembl"  "csavignyi_gene_ensembl"
```

This code fragment connects to the European Bioinformatics Institute's (EBI) Ensembl database and lists the names of the first few data sets held in that database. As we can see from this list, a list that actually extends to sixty-three species, there are many data sets in Ensembl. At row 8, hsapiens _gene_ensembl_ is likely to be a widely used one. What does it provide?

```
mart=useMart("ensembl", dataset="hsapiens_gene_ensembl")
listAttributes(mart)[1:10, 1]
[1] "ensembl_gene_id" "ensembl_transcript_id"
[3] "ensembl_peptide_id" "ensembl_exon_id"
[5] "description" "chromosome_name"
[7] "start_position" "end_position"
```

```
[9] "strand" "band"
g=getGene(id="100", type="entrezgene", mart=mart)
show(g)
entrezgene hgnc_symbol description
1 100 ADA adenosine deaminase [Source:HGNC Symbol;Acc:186]
2 100 ADA adenosine deaminase [Source:HGNC Symbol;Acc:186]
chromosome_name band strand start_position end_position
1 20 q13.12 -1 43248163 43280874
2 LRG_16 1 5001 37214
ensembl_gene_id
1 ENSG00000196839
2 LRG_16
```

Two things happen here. Some of the attributes of the hsapiens_gene_en sembl_ data set are listed, and the code then retrieves the database record for a gene with id=100 in the database, but more commonly known as ADA. The Ensembl record for ADA shows some information about the chromosomal location of the gene, its position in the human genome sequence (43248163, for instance, gives the base pair number at which the ADA starts in the full human genome sequence).

These brief code snippets and vignettes demonstrate not only the accessibility of sequence-related data but also some of the ways in which related data can be aggregated or superimposed. In a few lines of code it is possible to move across many databases and quickly move to almost any level of detail ranging from collections of genomes down to short sequences of DNA. The accessibility of sequence data sets is such that even a social science researcher can quickly write programs to retrieve these data. It attests to several decades' worth, if not longer, of work on databases, web and network infrastructures, and analytical software, all, almost without exception, driven by the desire for aggregation, integration, archiving, and annotation of sequence data that first became highly visible in the Human Genome Project (HGP) of the 1990s. The brevity of these lines of code—half a dozen statements in R, no more—suggests that we are dealing with a highly sedimented set of practices, not something that has to be laboriously articulated, configured, or artificed. Code brevity almost always signposts highly trafficked routes in contemporary network cultures. Without describing in any great detail the topography of databases, protocols, and standards woven by and weaving through bioinformatics, these examples suggest that the mixed, dirty, wide data sets fed to algorithms such as RF-ACE depend on the couplings and congregations of different databases connected into metadatabases such as Ensembl. As the code shows, sequence and other genomic data (and we will see some other types of contemporary genomic data below) are available not only to scientists as users searching for something in particular and retrieving specific data but also to scientists as programmers developing ways of connecting up, gathering, and integrating many different data points to produce the wide (many-columned), mixed (different types of data), and dirty (missing data, data that are "noisy") data sets digestible by RF-ACE. Somewhat recursively, corralling sequence data has led to highly developed genomic data infrastructures that come to serve almost as a kind of biological data commons whose primary resources—base sequences—are increasingly woven together with many other forms of comparison, searching, classifying, and sorting. The interwoven assemblage of genomic and biological databases, archives, software, and data standards is perhaps unrivalled by any other science in its variety and sheer density.

And again, the generic connectivity of these practices is not specific to genomics (although it plays out in specific ways there, affecting, for instance, the patterns of authorship, collaboration, and reputation in this field; see chap. 7 of this volume). The same sedimented, highly compressed layering of infrastructures, in which many details about protocols, transactions, architectures, and interfaces withdraw from view, can be found in many settings (in business, social media, supply chain management, etc.).

Forests and Lassos in Genomics

Highly leveraged infrastructures for access to biological data reshape what counts as a genome. As a data form, genomes in many ways become less linear or flat than the base sequences. The linear sequences of DNA data become more mixed and wide partly through the accessibility we have just seen that allows them to be superimposed, annotated, and layered. But their shape also changes for a different reason. A recent review in the journal *Genomics* highlights the increasing importance of the random forest (RF) machine learning techniques we saw in use during the Google I/O keynote: "High-throughput genomic technologies, including gene expression microarray, single nucleotide polymorphism (SNP) array, microRNA array, RNA-seq, ChIP-seq, and whole genome sequencing, are powerful tools that have dramatically changed the landscape of biological research. At the same time, large-scale genomic data present significant challenges for statistical and bioinformatic data analysis as the high dimensionality of genomic features makes the classical regression framework no longer feasible. As well, the highly correlated structure of genomic data violates the independent assumption required by standard statistical models."[8]

This kind of commentary on the changing shape, not just the volume, of genomic data is quite common. First of all, newer instruments or tools such as microarrays and faster sequencers (so-called next-generation sequencers) loom large. The tropes of waves, deluges, floods, and waves of sequence data being somewhat washed out, this account instead highlights the "high dimensionality of genomic features" and the "highly correlated structures of genomic data." The new ways of working with sequence data typically highlight changes in statistical or modeling approaches. So, for instance, Xi Chen and coauthors recommend the use of the RF algorithm because it "is highly data adaptive, applies to 'large p, small n' problems, and is able to account for correlation as well as interactions among features. This makes RF particularly appealing for high-dimensional genomic data analysis. In this article, we systematically review the applications and recent progresses of RF for genomic data, including prediction and classification, variable selection, pathway analysis, genetic association and epistasis detection, and unsupervised learning."[9]

The key terms on the machine learning side of this formulation are "large p, small n," "high-dimensional," "prediction," "classification," "variable selection," and "unsupervised learning." While these terms have been widely used in machine learning research since the early 1990s, they are becoming increasingly visible in genomics. The key terms on the genomics side of this formulation would perhaps be "pathway analysis," "genetic association," and "epistasis." These biological terms point to forms of relationality typically associated with biologically interesting processes. Epistasis, for instance, broadly refers to linked gene action, a process that has been difficult to study before high-throughput methods of functional genomics were developed. In contemporary genomic science, these biological processes are increasingly understood in terms of eliciting and modeling the relations between *features* of genomic data sets in order to classify and predict biological outcomes. In between the machine learning and the genomic references appear several statistical terms: "correlation" and "interaction." How does machine learning differ from the statistical practice that has underpinned much of modern biology?

The importance of statistics in life sciences is hardly new. The historian of biology Bruno Strasser suggests that an "important novelty of contemporary data-driven science is the omnipresence of statistical methods."[10] In commenting on the development of "big data" or "discovery mode" genomic science, Strasser presents episodes from the history of biology that prefigure contemporary turns to data as fount of epistemic value. He argues, for instance, that natural history practices entailed large repositories and archives of data-rich specimens and samples whose ordering and relations were constantly revisited and reassessed in the pursuit of scientific discovery: "natural history had been 'data-driven' for many centuries before the proponents of post-genomics approaches and systems biology began to claim the radical novelty of their methods. As I have argued here, many of what are claimed as novel features of contemporary data-driven science have parallels among earlier natural history practices."[11] As Strasser points out, the novelty of genomics lies not in collections of data that are interrogated for hitherto-unseen patterns of relations, but rather in the "omnipresence of statistical methods." In Strasser's account of the precedents for genomics, much hinges on what is meant by "statistical methods." I would suggest that genomics data have increasingly troubled mainstream statistical practices, and this trouble has led to many significant shifts in statistical technique since the 1990s. Among these shifts, machine learning represents a kind of extreme or brutal solution. But whether it is brutal or not, it cannot be simply seen as a continuation of statistical methods.

Some of the difficulties in statistical practice can be illustrated by genome-wide association studies (GWAS; note the resonance again of the term "wide"), studies that compare selected variations in the genomes of individuals from a given population. Note that GWAS studies rely heavily on the accumulated, augmented genomic data sets described previously. And GWAS, in turn, provide, for instance, the knowledge base on which 23andMe, the consumer genetic profiling service, draws. Typically, GWAS use microarrays to look for statistically significant associations between the presence of SNPs in individuals and the occurrence of diseases. While microarrays in many different forms are heavily used in research, they have been surprisingly slow to affect clinical practice. In their account of the surprisingly slow shift of microarrays toward clinical practice, Peter Keating and Alberto Cambrosio identify the rate-determining role of statistics: "The handling and processing of the massive data generated by microarrays has made bioinformatics a must, but has not exempted the domain from becoming answerable to statistical requirements. The centrality of statistical analysis emerged diachronically, as the field moved into the clinical domain, and is re-specified synchronically depending on the kind of experiments one carries out."[12]

Confronted by the wide data sets produced by microarrays, GWAS could only become answerable to statistics (that is, meet the standards of validity) by developing novel statistical practices, and here machine learning has played an important role. While Keating and Cambrosio interestingly suggest that statistical and bioinformatics practices are hybridizing in the development of GWAS, their analysis pays less attention to actual shifts in statistical practice coming into play in GWAS and many other areas of genomics. GWAS and microarrays are a case, perhaps quite low profile in some ways, in which machine learning techniques have pervasively affected genomics.

The problem is, as mentioned already, the dimensionality of data sets in GWAS. This dimensionality obstructs mainstream statistical practice. Again, as is often the case in genomics, thousands of these data sets are readily accessible in public databases such as the EBI's ArrayExpress and the National Center for Biotechnology Information's (NCBI) Gene Expression Omnibus (GEO).

If we query ArrayExpress for data sets relating to pneumonia in humans, forty-one are shown. (Most of these, we can see from the IDs, were deposited at NCBI's GEO but have been subsequently mirrored by ArrayExpress; mirroring is a crucial practice in the genomic infrastructures). Retrieving the processed, not the raw, data for one of them chosen at random:

```
Experiment data
Experimenter name: Woodruff, Prescott
Laboratory: UCSF
Contact information: prescott.woodruff@ucsf.edu
Title: Sarcoidosis-specific markers from whole blood gene expression
URL:
PMIDs:
No abstract available.
notes:
accession:
E-GEOD-19314
identifier:
E-GEOD-19314
experimentalFactor:
batch
years
diagnosis
dx
ethnicity
sex
type:
```

Some of the various types of "feature data" are listed. But the "high-dimensional" character—the "high p, small n"—of the data can be seen in an excerpt from the metadata:

```
exprs
Features 54675
Samples 66
[1] "1007_s_at" "1053_at" "117_at" "121_at" "1255_g_at"
[6] "1294_at" "1316_at" "1320_at" "1405_i_at" "1431_at"
```

The data sets produced by such studies are wide rather than long because they have *many more* features ($p=54,675$) than samples ($n=66$). In the context of GWAS, most of the features are SNPs. While a large study may have thousands of rows corresponding to tissue samples (for example, the Wellcome Case Control Consortium studied 14,000 individuals), it will certainly have many more columns since each SNP tested on the microarray will have its own column in the data set (up to a million or so).[13] A relatively small number of the columns of a typical GWAS data set often record the incidence of some biomarker (levels of blood lipids, etc.) or physiological variable (for instance, blood pressure, sex, or height). The vast bulk of the columns in a GWAS data set, however, will contain data from the microarray that points to the presence or absence of single mutations in the base sequence:

```
GSM479982_sample_table GSM479981_sample_table
1007_s_at 4.833 5.121
1053_at 6.127 6.180
117_at 10.246 11.205
121_at 5.657 6.068
1255_g_at 2.215 2.418
1294_at 8.792 8.151
```

Viewed from the standpoint of mainstream statistics, each cell on a microarray is effectively a separate assay or experiment, subject to its own statistical tests and inferences. Bringing together the analysis of patterns of distributions of hundreds of thousands of SNPs in GWAS and their association with disease represents a significant statistical problem in juggling statistical hypotheses. Finding the association between a *single* SNP and the disease or some phenotypical trait is quite straightforwardly done using standard statistical tests. But in GWAS, scientists do not know in advance which of the often several hundred thousand SNPs are associated with the trait. Furthermore, it is very unlikely that a single SNP is associated with a disease or given biological trait (for example, faster growth of a particular strain of an agricultural plant) in any case, since that would suggest that we are dealing with the relatively uncommon case of a single-gene trait or single-gene disorder. Much more likely, a set of SNPs scattered across the genome will be associated with the trait under study.

In statistical terms, GWAS analysis means testing hundreds of thousands of hypotheses simultaneously and deciding which of the associations between SNPs are likely to occur by chance. GWAS publications typically use Manhattan plots to highlight the associations between the trait or disorder and the occurrence of SNPs in sample genomes. The plots show how likely particular SNPs are to occur by chance in the presence of the trait or disorder. The p-values become smaller, and hence the ($-\log_{10}$) "buildings" on the Manhattan skyline become taller when SNPs are unlikely to occur by chance. The plot shows, in other words, the statistical significance of the occurrence of SNPs. The SNP at 19q13 has a p-value of around 10^{-15}, a vanishingly small probability of happening by chance. While there has been much debate about the proper ranges of p-values in GWAS, and the statistical treatment of GWAS has become increasingly sophisticated as analysis techniques take into the account the problems of false positives, these statistical developments are not to my mind the most significant shift in statistical practice associated with genomes as such.

At a genomic level, the main challenge lies not in the association of sets of SNPs with traits as such, but in working out which combination of the hundreds of thousands of associations between SNPs and traits is significant. Take, for instance, the task of analyzing epistasis or gene interaction in a GWAS. In epistasis, the level of expression of one gene is affected by its interactions with other genes. This would be an instance of a "highly correlated structure" in genomic data. The expectation is that analyzing how combinations of SNPs are associated with traits or disorders will yield insights into epistatic processes associated with disease or development. As a review of statistical reasoning in GWAS points out, "SNPs that combine to make larger genetic effects can statistically reflect an epistatic interaction, where the alleles of one gene influence the effects of alleles of another on a trait value or risk of disease. These interactions are by definition nonlinear and thus can dramatically increase the trait or risk."[14] Interaction between genes affects the expression of specific genes. This is seen to play an important biological role in complex disorders and also in development. Yet statistically modeling these interactions is, even in the eyes of statistically minded genomic researchers, extremely difficult. As Rita Cantor and co-

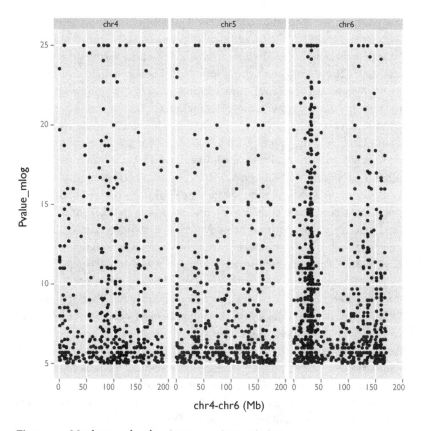

Figure 5.1. Manhattan plot showing SNP-trait associations.

authors point out, "In GWAS, even for main effects, the number of predictors far exceeds the number of observations. Exhaustive examination of all pairwise interactions is possible, but for multiway interactions the task is totally impractical."[15] The number of predictors or "features" is given by the number of SNPs analyzed in the study. This is currently around one million, suggesting an astronomical number of potential interactions. The network of possible interactions, as well as the combinatorial explosion of possible pathways between parts of genomes involved in these interactions, dwarfs readily available computational power.

Finding and statistically validating the possible association of a set of SNPs chosen from hundreds of thousands of variables requires a different approach to significance testing and statistical modeling. Cantor and coauthors write, "Recent advances in data mining . . . can be used to prioritize GWAS results."[16] What Cantor calls "data mining" is commonly called "machine learning" or sometimes "statistical machine learning" in order to emphasize the altered practices of inference it brings to bear on this problem.[17] Automated statistical model fitting—machine learning—promises to locate epistasis or other biological functions in the data sets produced by GWAS when standard statistical inference would wander aimlessly because of its difficulty in dealing with interactions or dependent processes.

In standard statistical inference, linear regression is often used to explore the relationship between variables and some outcome. In GWAS, the variables would be SNPs, and the outcome might be presence of disease or not. Linear regression and classification models are a backbone of statistical practices. The intuition behind them is drawing a straight line that best fits the data. Linear models "work" because normal statistical studies have many observations of a small number of "features" or "predictors," but as T. T. Wu et al. point out, "with hundreds of thousands of predictors, the standard methods of multivariate regression break down."[18]

The shape of GWAS data sets, with many variables (~500,000–1,000,000) and relatively few observations (100–5,000), means that linear regression cannot easily predict the best weighting of the variables. Faced with very wide data sets, the standard techniques of statistical analysis break down and have been increasingly supplanted by statistical machine learning techniques. The statistical techniques used in machine learning are somewhat different from mainstream statistical tests, modeling, and inference. Just like GWAS experiments themselves in their use of microarrays to effectively conduct many experimental tests at the same time, statistical machine learning usually constructs many models at the same time in order to then choose the most predictive model. (We have already glimpsed this with the RF approach, which constructs many decision trees and averages their results in order to classify or predict results.)

The essence of many of the machine learning approaches consists in optimizing hyperparameters. Machine learning techniques often find the optimum predictive combination of predictors (SNPs in this case) for the observed traits (disease, for instance), not by developing any kind of model of the phenomena under investigation. As in many other situations where machine learning is used (business analytics, image processing, etc.), the modeling techniques used are themselves somewhat agnostic as to biological explanation. (I return to this point in the conclusion: what does the generic character of machine learning imply for genomics?) Tech-

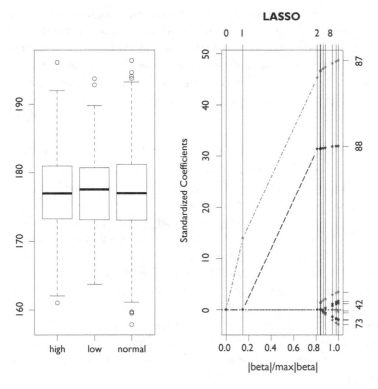

Figure 5.2. Lasso analysis of blood pressure SNPs.

niques and algorithms such as *lasso penalized regression*, *neural networks*, *logic regression*, and *Bayesian partitioning* that have been used to optimize the predictive mix of variables amid the myriad of possible associations know nothing of epistasis.[19] Instead of trying to track biological processes or biochemical pathways, they trace structures present in the data.

The somewhat daunting model *optimization function* for one commonly used machine learning technique called "lasso regression" is shown below, and this mathematical function displays some relevant features that might help us grasp how genomic data are handled in statistical machine learning:[20]

$$\hat{\beta}^{lasso} = argmin_{\beta} \sum_{i}^{N}\left(y_i - \beta_0 - \sum_{j=1}^{p} x_{ij}\beta_j \right)^2$$

subject to $\Sigma_{j=1}^{p}|\beta_j| \leq t$.

As in the case of the code-data vignettes for sequence database access, such mathematical expressions imply concrete operations that reshape genomic data. The forest of subscripts, letters, and mathematical functions is hard for social scientists and humanities scholars to look at, but these expressions abound in the statistical machine learning literature and litter the supplementary files of many genomic science publications. We do not need to understand them in great detail, but we do need to see in them some distilled expressions of things done to genomic data. For present purposes, perhaps the most outstanding features of the lasso estimate function are the two Σ (sum) operators that suggest that the function is repeatedly adding values. The inner Σ effectively generates a whole series of standard linear regression models that express outcomes as a linear function of combinations of predictors. The outer Σ creates an overall value for each model (using "least sum of squares"), but all the while shaping the important predictor coefficients β_j according to the second line of the model "subject to. . . ." This function seeks the minimum value (the argmin) of the hyperparameter β^{lasso} and at the same time constrains or "lassos" the number of predictors that appear in the model: lasso does a kind of "continuous subset [of features] selection."[21] Lowering the lasso variable t effectively shrinks the number of features (SNPs in this case) that appear in the model of the epistatic interactions: "making t sufficiently small will cause some of the coefficients to be exactly zero."[22]

In some ways, the lasso technique does not radically break with mainstream statistics. It is a way of choosing one linear regression model among many. Like all linear regression models, lasso is still trying to fit the best line to the data (or, actually, to a hyperplane). At the time, it reconstructs statistical practice fairly profoundly. Not only does lasso sift through many possible interactions in order to highlight in a model those combinations or features with higher predictive potential, but methods such as lasso also seek to identify the most *sparse* model possible. A sparse linear model, one that minimizes the number of features, facilitates biological interpretation. Given a smaller number of significant predictors, biological scientists can then examine some combination of variables or predictors more closely, perhaps using laboratory techniques or through further integrative data analysis, in order to track down the pathways or the mechanisms through which the regions of DNA around the most highly predictive SNPs act biologically. In our example of analysis of epistasis, the most important feature of the machine learning techniques lies in the way that they seek to eliminate the highly correlated associations between SNPs that hold little predictive value (for instance, because they are clustered in one locus on a chromosome, and all relate to a single well-known gene) and highlight uncorrelated associations between SNPs that suggest hitherto-unexplored pathways of gene interaction.

Localizing Variation: 463,320 kNN and the Clinical Endpoints of Microarrays

The broader point here—I think it matters greatly for postgenomic science—is that the shift from one model to many models reorganizes biological sciences. The machine-learned models flow back through genomic science, reorganizing and refocusing research, experiments, databases, and funding in subtle but important ways. Between the pre-HGP and post-HGP science, the status of significant differences in genomes shifted. Pre-HGP biology understood the significant differences between individual organisms largely in terms of gene alleles responsible for variations in phenotypes. Biological differences, and disease in particular, stemmed from different forms of genes. Understanding disease meant finding the disease genes. Even prominent proponents of genomics, such as Leroy Hood, writing of "Biology and Medicine in the Twenty-First Century" in 1991, envisaged genomics as a way of simplifying "the task of finding disease genes."[23] Across the taxonomic hierarchies, genes were the object of much annotation, labeling, and description. GenBank, the primary repository of gene sequences, still embodies this conception of variation. Two decades after the inception of the HGP, genomes present a different image of variation. According to Nikolas Rose, writing more recently, "there is no normal human genome; variation is the norm."[24] "In this new configuration," he writes, "what is required is not a binary judgment of normality and pathology, but a constant modulation of the relations between biology and forms of life, in the light of genomic knowledge." The emphasis in Rose's formulation falls on "constant modulation" of the relations between biology and forms of life. If postgenomic science departs from the understanding that there is no single genome but many genomes (see also chap. 4 of this volume), then according to Rose, variation itself becomes of primary interest. Here I want to argue that pursuit of variation is remaking the genome into "a form whose only object is the inseparability of distinct variations."[25] These variations are, as we will see, located within genomics understood as a technical enterprise, between genomes and within the genome. In all three loci of variation, machine learning comes into play.

Within genomics itself, machine learning figures as a way of correcting for the fact that genomic instruments often produce different results, and that genomic knowledge claims shift unpredictably. That is, machine learning has increasingly been directed to the reorganization of genomics itself. For instance, since microarray data themselves suffer from many such problems of variation, the U.S. Food and Drug Administration has since 2003 conducted a study of data analysis techniques for microarrays: "The US Food and Drug Administration MicroArray Quality Control (MAQC) project is a community-wide effort to analyze the technical performance and practical use of emerging biomarker technologies (such as DNA microarrays, genome-wide association studies and next generation sequencing) for clinical application and risk/safety assessment."[26] (Keating and Cambrosio mention the MAQC consortium in their account of the hybridization of genomics and biostatistics.[27]) Phase I of the U.S. Food and Drug Administration–led MAQC addressed many issues of data analysis in the context of the clinical applications of gene expression analysis using microarrays. The primary statistical issue there was minimizing the "false discovery rate," a typical biostatistical problem.[28] In MAQC-II, however, the focus rested on the construction of predictive models for "toxicological and clinical endpoints . . . and the impact of different methods for analyzing GWAS data."[29] On both the clinical and GWAS fronts, the thirty-six participating research teams tried out many predictive classifier models. Different machine learning techniques generate different kinds of models; genomic and biomedical researchers are compelled to engage with the many variations in prediction.

In the shift from MAQC-I to MAQC-II, the problem of variations in the predictions produced by the machine learning models moved to center stage. The problem of variation arises not because any of the different modeling strategies used in machine learning gene expression data sets are wrong or erroneous, but because every model moves through the "feature space" in a different way.[30] In the MAQC-II consortium, the teams were tasked to build "classifiers," models that predict whether a given sample or case belongs to a "normal" or "disease" group. This is a typical machine learning task. On this point, the design and practice of MAQC studies, and indeed most GWAS, do not accord directly with Rose's suggestion that "constant modulation," not "binary judgment," is required. Classifiers often make binary judgments. Now, the most popular classifier in the MAQC consortium was the kNN model: "among the 19,779 classification models submitted by 36 teams, 9,742 were k-nearest neighbor-based (kNN-based) models (that is, 49.3% of the total)." However, these models varied greatly in their predictions: "there have been large variations in prediction performance among KNN models submitted by different teams."[31] Here we can begin to see that not only the genome but the models themselves vary. In seeking to track and predict genomic variation, postgenomic science has constructed a new zone of variations at the intersection of database infrastructures and sequencing or gene expression technologies.

As in many aspects of genomic sciences, a typical response to the problem of variation is to count things. Genomics is one of the greatest commitments to counting things ever undertaken. In their attempt to normalize the variations of popular predictive models, one of the research groups in MAQC-II writes that "for clinical end points and controls from breast cancer, neuroblastoma and multiple myeloma, we systematically generated 463,320 kNN [k-nearest neighbor] models by varying feature ranking method, number of features, distance metric, number of neighbors, vote weighting and decision threshold."[32] For present purposes, the striking feature here is the proliferation of models in an effort to tame the variations of predictive models. The number of predictive models constructed here rivals the number of SNPs assayed by the microarrays. Why do these models, in this case of exactly the same kind of data we have discussed above, multiply so greatly?

All data analysis faces the so-called curse of dimensionality,[33] but genomic data are particularly "cursed" by their high dimensionality, by what I have termed above their "expansiveness." Every distinct feature (e.g., a SNP) in a data set effectively adds a new dimension to the data space. If every SNP, or every RNA transcript, adds a new dimension in the feature space, efforts to model this feature encounter high-dimensional spaces. Now nearly all classifier models seek some kind of regularity or boundary that cuts the space into two or more regions (e.g., "diseased" and "normal"). The line or surface that separates data points in a classifier can be linear (as in the lasso models) or nonlinear (as in kNN models). Since the 1950s, problems of classification and prediction in high-dimensional data spaces have been the object of mathematical interest. The mathematician Richard Bellman coined the term "the curse of dimensionality" to describe how partitioning becomes more unstable as the dimensions of the data space increase.[34] The problem is that while the volume of a space increases exponentially with dimensions, the volume of data usually does not increase at the same rate. In high-dimensional spaces, the data become more thinly spread out. This makes it hard to construct good partitions. Sparsely populated spaces accommodate many different boundaries.

Most machine learning techniques embody some way of managing the dimensionality of the feature space, by either effectively reducing that dimensionality (as we saw in the lasso technique, which corrals the numbers of predictive features in the model) or concentrating on localized regions of the feature space, as does kNN. For instance, the kNN models analyzed by Parry as part of MAQC-II are widely used because they are the simplest way of dealing with interactions in an "exceedingly large feature space."[35] In kNN, a training set of observations whose outcomes are known (e.g., clinical endpoints in cancer might be benign or malignant) is used to help classify (and hence predict) the test data. For our purposes, the interest of this technique lies in its way of traveling through the data. Rather than trying to find the line of best fit (as, for instance, the linear regression models do), kNN models find local clusters in the data and create irregular boundaries in the data.

As we can see from the kNN model depicted in figure 5.3, in which data belong to two classes (normal vs. not normal), the decision boundaries produced by the algorithm can be unstable. The example shows two models: one for $k=5$, the other for $k=15$. Each model examines the relations between 5 and 15 points in deciding whether a particular case belongs to one class or another. While kNN' finds local clusters and classifies on the basis of an irregular decision boundary, this classificatory power comes at the cost of instability. As Hastie, Tibshirani, and Friedman put it, "The method of k-nearest neighbors makes very mild structural assumptions: its predictions are often accurate but can be unstable."[36] These shifting boundaries stem from the dimensionality of the data. The more dimensions or features in the data set, the larger the local neighborhood needed to capture a fraction of the volume of the data, and the more likely that most sample points will lie close to the boundary of the sample space, where they will be affected by the neighboring space. The result is that "in high dimensions all feasible training samples sparsely populate the input space."[37] Because kNN allows for nonlinear interactions between features, for instance, small differences in the number of points in particular neighborhoods can drastically affect the boundaries (as we see in comparing the right- and left-hand plots in figure 5.3). These kinds of topological instabilities account for the propensity of machine learning treatments of feature-rich genomic data to produce highly variable predictions. We can also see why the MAQC-II teams produced 463,000 kNN models in an effort to normalize and regulate predictive practice. The price of predictivity in genomics is variation in prediction.

5 - nearest neighbor

15 - nearest neighbor

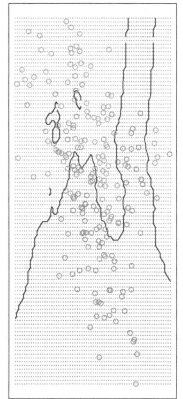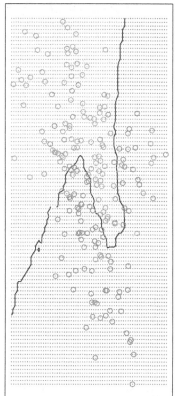

Figure 5.3. kNN model.

A Self-Organizing Map of Genomic States

If predictive models polymorphically proliferate in genomics, what of genomes themselves? Do biological understandings of the genome shift as a result? A final example, this time of a major genomic project, indicates what happens. The vast consortial structure of the ENCODE project focused on DNA outside the more well-known protein-coding regions of the human genome (the regions so extensively annotated in the millions of records in GenBank, for instance).[38] The architecture of this decade-long project ("a landmark in the understanding of the human genome," according to the *Nature* headline in September 2012) is too complicated to follow here, but machine learning in some form or other figures in a substantial portion of the thirty publications that marked the completion of the initial analysis of the project data. In an overview of the results, the ENCODE authors describe the thousands of data sets and hundreds of cell types they worked with in seeking to "delineate all functional elements encoded in the human genome."[39] The overview is replete with graphics that nearly all point to large amounts of data processing and statistical computation. Of the seventy-two discrete visual figures in the leading ENCODE article, only two are somewhat image based (stained blood vessels in a transgenic mouse embryo; a transgenic medaka fish). The rest are model-derived figures.

Whereas the HGP sought to develop over three five-year periods three different kinds of maps—genetic, physical, and sequence[40]—ENCODE presents a different kind of map, a *self-organizing map* (SOM) of genomic states (see figure 5.4). "Segmentation" of the human genome was a major concern in ENCODE. The term, not coincidentally, comes from data mining and marketing research rather than biology. Segmentation refers to a way of sorting "data into a small number of discrete states based on the continuous output values."[41] As Hastie et al. describe it, "Cluster analysis, also called data segmentation, has a variety of goals. All relate to grouping or segmenting a collection of objects into subsets or 'clusters,' such that those within each cluster are more closely related to one another than objects assigned to different clusters. An object can be described by a set of measurements, or by its relation to other objects. In addition, the goal is sometimes to arrange the clusters into a natural hierarchy."[42]

Segmentation is particularly important to ENCODE because it aims to encyclopedically describe the functional elements of the genome. The seven state types predicted by ENCODE, again using sophisticated prediction techniques (e.g., "Segway," a dynamic Bayesian network modeling tool),[43] refer to functionally differentiated sites on the genome (e.g., state E is an enhancer; state T is a predicted transcribed region).[44] Genomic segmentation is a major machine learning problem in its own right since it does not refer simply to sequences but to patterns of biochemical signal that occur on many different scales, but here I am less interested in the different states attributed to genome than in the way that the predictive models that segment the genome begin to change what a genome is. Using segmentation as the basis of its description, ENCODE begins to reconceptualize genomic "function" as segmentation.

The high-resolution segmentation of ENCODE data using self-organizing maps (SOM) is a good example of what happens to genomes in machine learning. An SOM, as Hastie et al. tell us, is an unsupervised learning technique in which "the prototypes are encouraged to lie in a one- or two-dimensional manifold in the feature space. The resulting manifold is

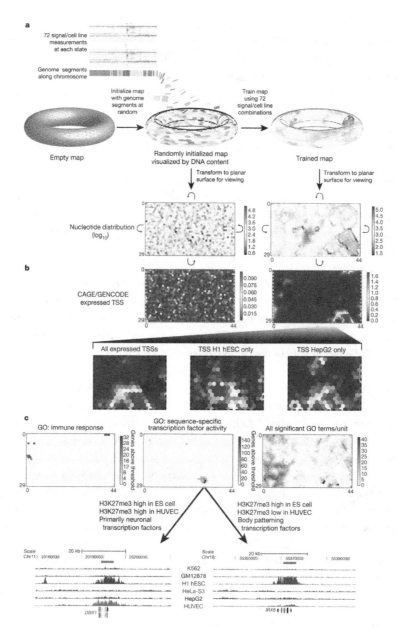

Figure 5.4. Training self-organizing maps. Credit: Reprinted by permission from Macmillan Publishers Ltd: ENCODE Project Consortium, "Integrated Encyclopedia of DNA Elements in the Human Genome," figure 7.

also referred to as a *constrained topological map*, since the original high-dimensional observations can be mapped down onto the two-dimensional coordinate system."[45] Leaving aside the mathematical and computational details of how this is done, we see here dimensional reduction that embeds a high-dimensional manifold in a low-dimensional space. Like many other figures in the ENCODE publications, the SOM of genomic states leverages a variety of data sets against each other. Figure 5.4 usefully points to the re-combinant transformations and transpositions of sequence data in genomics as machine learning techniques reshape it. Figure 5.4 itself comprises sixteen subfigures that display progression from genome segments (themselves already modeled, as discussed above) to training of the SOM using cell/signal measurements (shown as a toroid), and then to illustration of the results of the SOM as applied to some elements of the Gene Ontology, a major bioinformatics initiative aiming to standardize the naming of genes and gene attributes.

The ENCODE article displays the SOM as a toroid or doughnut, a manifold that has no edges. This toroidal map is initially randomly populated with all the segments found by the segmentation analysis. The ENCODE authors write, "We then trained the map using the signal of the 12 different ChIP-seq and DNA-seq assays in the six cell types analysed."[46] Once the training has been completed, "the resulting map can be overlaid with any class of ENCODE or other data to view the distribution of that data within this high-resolution segmentation."[47] They go on to show how other data sets can be overlaid on the map (see the last line of figure 5.4). In a recursive move that I would suggest is typical of the subductive force of machine learning in genomic research, the power of the SOM is demonstrated by reordering other genomic data sets, such as annotated sets of genes (GENCODE), or gene ontology (GO) data sets. In the final analysis, the toroidal SOM is used to again annotate regions on two tracks in genome browser-style visualizations of DNA. The output of the predictive model becomes a way of reorganizing existing maps and catalogues of genomic features: SOM-annotated regions of the genome can be displayed as tracks in a genome browser.

Genomic Landscapes: Open or Settled?

There are different ways of making sense of the implication of computation in genomic and postgenomic science. We might attend to the long-standing and still-operative metaphors of DNA as code, program, communication system, or network;[48] to the rich and diverse development of information retrieval and database systems dedicated to retrieving and annotating bio-logical data of many different kinds (the annual database issue of the jour-nal *Nucleic Acids Research* lists over four hundred); to the role of various kinds of system models in systems biology and dynamic models in biology more generally;[49] to the crucial role played by certain kinds of matching and alignment techniques in sequencing;[50] to the rise of bioinformatics as a bundle of techniques focused on cross-linking sequences and annota-tions; or, stretching further afield, to the attempts to reconfigure DNA and RNA sequences as components that function like digital logic in synthetic biology.

In the predictive models of epistasis in GWAS, in standardization of the predictions of clinical outcomes using kNN models in MAQC-II, and in the SOM of genomic states in ENCODE, we glimpse a set of transformations that are changing the shape of genomic data and thereby how genomes matter and what they mean. These examples, while by no means trivial, are inevitably quite limited in their scope. Yet they are not isolated. There are many other such examples. We could look at the large body of work that seeks to elicit subtypes in relation to diseases, disorders, or traits from populations.[51] We could examine the processes of imputation of genotypes that play a key role in firming up the statistical power of GWAS to identify associations.[52]

Machine learning widely affects epistemic infrastructures, techniques, and scientific knowledge. In the biosciences, it touches on those aspects of genomes—their variation, their manifold spatial and temporal relationality in biological processes—that seem most distant and difficult to derive from the relatively stable, monolithic, and hence tractable forms of order found in DNA sequences. As we have seen in the various data excerpts above, that order is largely linear. DNA can be laid down in tracks, aligned, and an-notated, even though it undergoes subtle, pervasive, and transient forms of temporal and spatial reshaping in life-forms. In the feature-rich spaces countenanced by machine learning, we see attempts to embed manifolds in local regions, local linearities. Sometimes these local regions are regions of annotated DNA, as in the ENCODE examples, or nonlinear interactions between sets of genes, as in the GWAS analysis of epistasis. At other times, these local regions are forms of life in a more general sense—clinical out-comes or diagnostic tests—as in MAQC-II.

I have emphasized on several occasions the common thread of what could be termed the "genomic curse of dimensionality." The genomic sci-ences are based around the extraction, sorting, and ordering of DNA se-quences. Machine learning practices, I have suggested, begin to construct different dimensional spaces amid the sequences, lines, and tracks of genomes. What perhaps most vexes and animates postgenomic science is the desire to separate out from the DNA sequences variations that matter to both life-forms and forms of life. This vexation generates strenuous infrastructural, technical, and conceptual attempts to reorganize and reconceptualize what it is to know a genome. The machine learning techniques I have discussed have begun to transform bioinformatics and biostatistics. To understand machine learning either as intensification of statistics, as hybridization of bioinformatics and statistics, or as the product of a reciprocal interactive convergence of biology and computing misses some of the important features of the reorganization of genomics associated with these practices. It misses what machine learning brings from elsewhere to genomics, and it overlooks how the genome itself serves as a showcase, as we saw in the Google I/O demonstration, of certain much wider transformations in the practices of anticipation and prediction.

What is at stake in machine learning for postgenomic science or practices of knowing and predicting more generally? Could we pose or address any normative questions by becoming aware of and articulating machine learning in practice with greater clarity? Genomic science, in what it borrows from and in how it is affected from machine learning, displays some of the tendencies to reduce divergences and to corral differences typical of knowledge economies more generally. The philosopher of science Isabelle Stengers writes, "With the knowledge economy, we may have scientists at work everywhere, producing facts with the speed that new sophisticated instruments make possible, but that the way those facts are interpreted will now mostly follow the landscape of settled interests. . . . We will more and more deal with instrumental knowledge."[53]

As we see in the 600,000 cores of Google Compute applied to exploration of associations in cancer genomics using RFs, or the lasso applied to microarray SNP data, machine learning rapidly produces facts. Stengers suggests that the risk here is that divergence and unexpected forms of experimental results are somewhat diminished as a result. Machine learning in genomics might produce a "self-organizing map" that poses questions following the "landscape of settled interests" or status quo.

The very justification for using such techniques is the inordinate difficulty of exploring the many dimensions of genomes otherwise. I have already hinted at what I think is one core stake in these shifts. Genomes and genomics are touchstones for wider transformations in many sectors of science, industry, commerce, media, and government susceptible to the imperatives of the knowledge economy. In contrast to some of these domains, where much that happens is obscured from view, the great virtue of genomic science is the relative openness of its workings and its dogged insistence on DNA as the generating set. The fact that data practices are relatively generic and accessible means that critical research into transformations associated with data and knowledge economies can accompany nearly every aspect of genomic practice. This is a forensic good.

Notes

1. Krzywinski et al., "Circos."
2. Google Inc., "Behind the Compute Engine Demo."
3. Anthony, "Google Compute Engine."
4. Google Inc., "Behind the Compute Engine Demo."
5. Tuv et al., "Feature Selection with Ensembles," 1341.
6. R Development Core Team, "R Project for Statistical Computing."
7. Fisher, "Use of Multiple Measurements in Taxonomic Problems."
8. Chen and Ishwaran, "Random Forests for Genomic Data Analysis," 323.
9. Ibid.
10. Strasser, "Data-Driven Sciences," 87.
11. Ibid.
12. Keating and Cambrosio, "Too Many Numbers," 49.
13. Burton et al., "Genome-Wide Association Study of 14,000 Cases."
14. Cantor, Lange, and Sinsheimer, "Prioritizing GWAS Results," 10.
15. Ibid., 11.
16. Ibid.
17. Hastie, Tibshirani, and Friedman, *Elements of Statistical Learning*.
18. Wu et al., "Genome-Wide Association Analysis," 714.
19. Cantor et al., "Prioritizing GWAS Results," 11.
20. Hastie et al., *Elements of Statistical Learning*, 68.
21. Ibid., 69.
22. Ibid.
23. Hood, "Biology and Medicine in the Twenty-First Century," 138.
24. N. Rose, "Normality and Pathology in a Biomedical Age," 75.
25. Deleuze and Guattari, *What Is Philosophy?*, 21.
26. Parry et al., "K-Nearest Neighbor Models," 292.
27. Keating and Cambrosio, "Too Many Numbers."
28. Slikker, "Of Genomics and Bioinformatics," S1.
29. Ibid., 246.
30. Parry et al., "K-Nearest Neighbor Models," 292.
31. Ibid., 293.
32. Ibid., 292.
33. Hastie et al., *Elements of Statistical Learning*, 22.
34. Bellman, *Dynamic Programming*.

35. Parry et al., "K-Nearest Neighbor Models," 292.

36. Hastie et al., *Elements of Statistical Learning*, 11.

37. Ibid., 23.

38. ENCODE Project Consortium, "ENCODE (ENCyclopedia Of DNA Elements) Project," "Identification and Analysis of Functional Elements," and "Integrated Encyclopedia of DNA Elements."

39. ENCODE Project Consortium, "Integrated Encyclopedia of DNA Elements," 57.

40. Hood, "Biology and Medicine in the Twenty-First Century," 137.

41. Day et al., "Unsupervised Segmentation of Continuous Genomic Data," 1424.

42. Hastie et al., *Elements of Statistical Learning*, 501.

43. See Hoffman et al., "Unsupervised Pattern Discovery."

44. For summary of the states, see table 3 and fig. 5 in ENCODE Project Consortium, "Integrated Encyclopedia of DNA Elements."

45. Hastie et al., *Elements of Statistical Learning*, 528.

46. ENCODE Project Consortium, "Integrated Encyclopedia of DNA Elements," 67.

47. Ibid.

48. Kay, *Who Wrote the Book of Life?*

49. Szallasi, Stelling, and Periwal, *System Modeling in Cellular Biology*.

50. Stevens, "Coding Sequences."

51. See, e.g., Watson, *Recombinant DNA*, 3.

52. See, e.g., Burton et al., "Genome-Wide Association Study of 14,000 Cases."

53. Stengers, "Experimenting with *What Is Philosophy?*," 377.

6

Networks

REPRESENTATIONS AND TOOLS IN POSTGENOMICS

Hallam Stevens

The Network

A cursory glance over recent literature in the life sciences indicates that "networks" are becoming critical tools for understanding organisms. Biologists talk about gene regulatory networks, protein interaction networks, cell-signaling networks, metabolic networks, and ecological networks. The dictionary definition of a network includes any set of interconnected objects: cities might be connected by a railway or bus network, people are connected by mutual professional or personal contacts into a social network, or the various locations of a business could form a network of branches. More formally, the objects are "nodes," and the connections between them are called "edges." The term "network" is usually reserved for situations involving a relatively large group of nodes connected by an even larger number of edges to form a complex "web" of connections.

In the early twenty-first century we are thinking about particular kinds of networks all the time. Many of us spend most of our time connected to computer networks or cell phone networks, and our lives are dominated by social and professional networking forums such as Facebook and LinkedIn. Some of these networks are more tangible (based on wired connections, roads, or railway tracks), while others are more abstract (such as the network of our friends and acquaintances). However, the idea of a network allows us to think about all these phenomena as somehow similar; even though our social network does not literally consist of physical links between all of our friends, we imagine and talk about it as having many of the properties of a physical network. In this sense, the idea of the network conjures like objects from remarkably different media: we understand one conceptual domain (our friends, for example) in terms of another (a set of physically linked objects).

Biological networks, like cell phone networks or social networks, do not consist of stable physical links between fixed entities. The objects in a biological network (pieces of sequence or proteins, for instance) are not permanently linked in any way—their "interconnection" consists of transient spatial interactions.[1] Second, the objects that are supposed to be connected are not always the same objects. When we say that protein *a* interacts with protein *b* and protein *a* also interacts with protein *c*, the objects corresponding to protein *a* in each case need not be the same (they are usually physically different molecules).[2] The idea of a "biological network" is therefore a kind of fiction: it is a way of thinking about biology as an interconnected system of objects. The aim of this chapter is to explore the origins and consequences of this reimagining.

From Metaphors to Representations and Tools

Science studies has done much to show how metaphors influence the research agendas scientists adopt, the questions they ask, the models they use, the theories they develop, and the experiments they do. In short, metaphors are constitutive of how scientists understand the world. Words and the descriptions we make with them influence scientific thought. Metaphorical descriptions make some questions, problems, and solutions obvious, while making others invisible. Metaphors are productive: they suggest new areas of research or new ways of tackling a problem or understanding a system. But they also constrain, locking in particular ways of thinking and doing.[3] Describing the interior of an atom in terms of the solar system, for instance, suggests the possibility of measuring the speeds of orbiting electrons and their distances from the central nucleus. But it simultaneously closes off the possibility of understanding electrons as waves in an electromagnetic field.

Molecular biology has been dominated by the metaphor of information. Biologists such as G. C. Williams and Richard Dawkins worked to articulate the consequences of understanding DNA and RNA as informational entities.[4] The work of historians Lily Kay and Evelyn Fox Keller in particular has shown how the relationships between DNA, RNA, proteins, and organismic bodies have been constituted in terms of the flux of information.[5] The rise of computers, cybernetics, information theory, and communication theory alongside molecular biology allowed biologists to understand their molecules in terms of "codes," "scripts," and "texts." DNA became—through metaphor—a "master molecule," a kind of blueprint from which the organism could be built and operated.

The most profound consequence of the information metaphor was that it reinforced genetic determinism and sequence reductionism. According to the well-documented story, it promoted a model of life in which DNA-as-code-script remained at the center. Epigenetic, developmental, cytogenetic, and environmental approaches were pushed into the background.[6] Central to this metaphor is the image of the computer as an information-processing machine. The computer allowed biologists to imagine DNA as a software program that read off and executed the plan for an organism.

The growing importance of networks suggests a possible alternative to the long-dominant code-script. Networks, however, are more than just texts. First, they are *visual*—they are powerful ways of representing complex systems in two-dimensional space. The fact that train networks, social networks, and biological networks can be visually represented in exactly the same way gives network thinking much of its power. The very same diagram (see figure 6.1) might represent a set of computers, friends, or proteins. Second, network science involves a distinct set of *mathematical tools*. Imagining a collection of objects as a network allows these tools to be applied to it. What links train networks, social networks, and biological networks is not merely the possibility of their equivalent representation, but also that they thereby become susceptible to an equivalent set of mathematical and scientific practices.

It is not just words or descriptions that are important in guiding scientific thinking. Images are important too. And the usefulness of specific tools, too, can cause scientists to perceive objects and sets of objects in particular ways. Visualizations and other tools critically shape scientific knowledge;[7] network science provides ways of seeing, thinking, and doing that are now crucial to the way many biologists think about organisms. Like metaphors, representations and tools make some problems and solutions obvious and others invisible. The fact that certain visualizations and tools are ready-to-hand (or ubiquitous or easy to use) influences how scientists choose to represent and manipulate their objects of study.

The representations and tools of network science have begun to fundamentally shape how we see biological objects. In the 1990s, the Internet and World Wide Web emerged as new kinds of machines for engineers, physicists, and biologists to think with. During that decade, scientists and social scientists struggled to measure, describe, analyze, and understand online networks, developing and bringing together an armory of tools. Around 2000, these tools began to be deployed in molecular biology and genomics. The Internet, the World Wide Web, and social networks became objects with which to do biology. Now, networks increasingly constrain how we see and understand life.

This does not mean that the information metaphor has disappeared. On the contrary, the information metaphor appears to be as powerful as ever. The massive amount of genomic sequence data, which continues to be accumulated, attests to the continuing centrality of DNA. The information metaphor described the computer as a master machine. But the computer is a polyvalent entity, a protean machine: it offers up multiple metaphors. Keller has suggested that already in the 1970s the computer was a device "based not on unidirectional transmission of messages from sender to receiver but on networks and systems."[8] The tools and representations of biological networks can rest comfortably on top of biology as information. After all, it is information that flows through computer networks too. Many biologists are committed to *both* the centrality of DNA and the importance of networks—these are not mutually exclusive views of the biological world.

What kind of an account of life do networks offer us? First, network science has increasingly made *nonbiological* explanations of biological functions plausible. After presenting a review of the emergence of network science in the next section, I turn to a discussion of these new kinds of accounts in the following section. Second, networks draw into view the interconnections and interactions between biological parts, and especially parts of genomes. In particular, networks have made accounts based on the *structure* or *wiring* of sets of biological parts possible and even powerful. Some examples of this are given in the penultimate section. In the concluding section I suggest some of the implications of network thinking for complexity, holism, and reductionism.

Network Studies

Some time in the early 2000s, a field calling itself "network science" emerged. This discipline provided a well-developed set of tools, models, and representational practices for conceptualizing and studying diverse phenomena, including biological ones. This formalism provided the basis for the use of networks in biology. This section describes the emergence of "network science" out of various strands of work in the natural and social sciences. Despite practitioners' claims for the novelty of this discipline, it is important to note that "networks" of various sorts have long played a significant role in thinking about bodies and organisms. In the nineteenth century, the nervous system was understood in terms of electrical signals using telegraphic systems as an explicit model.[9] After World War II, cybernetics (the science of feedback systems) also imagined organisms in terms of networks of circuits.[10] Building on this work, Ludwig von Bertalanffy and others developed "systems theory" as a set of mathematical and analytical techniques for perceiving similarities between elements organized together in different contexts.[11]

In the 1980s, systems science proliferated and branched into a variety of new fields: catastrophe theory, context theory, dynamical systems theory, and complexity theory, among others. Many of these brought an increasingly sophisticated mathematical approach, especially the use of chaos theory and nonlinear dynamics for modeling. The Santa Fe Institute, established by a group of physicists and mathematicians in 1984, concentrated on the study of "complex adaptive systems." These included economic systems (the stock market), social insects, the biosphere, the brain, the immune system, the cell, and social systems such as political parties and communities. Such interdisciplinary work was motivated by the fact that all these objects were taken to have fundamental properties and behaviors in common: they were systems of elements that could change, adapt, and learn.[12]

Despite the prestige of Santa Fe, systems approaches in biology remained marginalized within the discipline as a whole. Complex adaptive systems were associated more with the study of artificial life (that is, life simulated on a computer) than with real organisms. The mathematical and computational bent of systems theoretic approaches, too, kept them separated from a discipline more comfortable in the laboratory and historically averse to "theoretical" approaches.[13] Only after the growth of the World Wide Web and the emergence of network science did biology begin to embrace some of the tools and ideas of systems theory.

During the early 1990s, the invention of the World Wide Web and the graphical web browser (as well as investment in infrastructure by public and private organizations) caused exponential growth of the Internet. The Internet became a mass phenomenon, with connections quickly becoming ubiquitous in business, scientific, academic, and domestic contexts.[14] The rise of the web generated a groundswell of scientific interest in networks and their effects. The Internet became the exemplar for studying networks. Understanding the growth and dynamics of the Internet itself became an

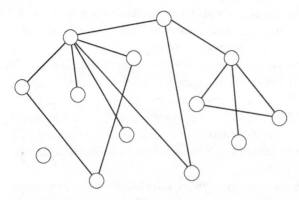

Figure 6.1. A typical depiction of a network. Circles represent nodes, and lines represent the edges joining the nodes. These nodes could represent cities in a rail network, people in a social network, computers connected in an electronic network, or proteins that interact with one another (a protein interaction network).

important undertaking. Economists and business analysts in the midst of the dot-com boom wanted to understand how "network effects" drove market dynamics. Sociologists (and advertisers) wanted to understand the effects of e-mail, online chatting, and (eventually) social networking on communities and societies. Media theorists wanted to understand the effects of hypertext writing. Political scientists wanted to understand the effects of decentralized information flow on democracy. All these tasks required paying attention to the Internet as a network object.

The tools developed for these analyses drew on the systems theory tradition. But they also added some new elements. Most importantly, network studies drew on the mathematics of graph theory. Most histories of graph theory locate its origin in Leonhard Euler's famous treatment of the "Seven Bridges of Königsberg" in 1736.[15] Graph theory is now a branch of topology that studies the connections between objects in a collection. The objects are modeled as "vertices" (nodes or points) and the links as "edges" (lines between points).[16] The result looks something like a flowchart (see figure 6.1). Modeling the Internet (and other networks) as graphs (with computers as vertices and wires between them as edges) allowed scientists to bring a range of new mathematical tools to bear on networks.

The importance of the Internet to the development of network science is demonstrated by Albert-László Barabási's route into the field (Barabási is widely considered to be one of the founders and leaders of network science). In 1998, he asked one of his students to undertake a side project from the lab's usual work (Barabási's lab worked on the material sciences, especially the physics of rough surfaces).[17] The project was to map out some of the connections on the World Wide Web. The result was a surprise: instead of links being randomly distributed, they appeared to follow a power-law distribution. This meant that most web pages were linked to only a few other pages, while a small number had a large number of links. Barabási and his students looked at other networks, including transistors in a computer chip and actors in movies. The power law appeared there too. Barabási (working with his student Réka Albert) realized that such a network could be generated by "preferential attachment"—that is, by adding connections preferentially to those nodes that already had lots of connections.[18] Their model of "scale-free" networks, as they called them, was published in *Science* in 1999.[19]

The new network studies also drew on the sociology and anthropology of "social network analysis." Network science cites work by Anatol Rapoport on social networks and the spread of information, Mark Granovetter on weak ties, Stanley Milgram on the small-world problem, and Herbert Simon on social groups.[20] This work in mathematical sociology, in turn, was originally linked to cybernetics (Rapoport, for instance, cofounded the Society for General Systems Research with Bertalanffy, Ralph Gerard, and Kenneth Boulding in 1954) and mathematical biology (particularly the work of Nikolas Rashevsky).[21] In the late 1990s, network scientists began to apply these sociological concepts to the analysis of computer and other networks. Bringing these together with the mathematics of graph theory, they either invented or revived a set of tools for characterizing networks. These included measures of size, density, average degree, diameter, clustering, robustness, and centrality, as well as concepts such as scale-free, preferential attachment, percolation, link analysis, and associative mixing.

Again, the Internet—increasingly understood as a *social* as well as an electronic network—provided the inspiration for these new kinds of analysis. The work of physicist Duncan Watts, another network studies pioneer, has found its most important applications in understanding online social networks. In 1998, Watts, with his PhD supervisor Steven Strogatz, published a mathematical model of "small-world" social networks.[22] This provided him with a powerful research tool that he applied to the analysis of crowd behavior, epidemics, public opinion formation, trends in social networks, cooperation, and information exchange in organizations. Abstracted as networks, all of these phenomena could be analyzed and understood on the same terms.

By the mid-2000s, network science was a booming new interdisciplinary field. By 2005, papers by Barabási and Albert and by Watts and Strogatz had been cited more than 1,500 times each. Between 2000 and 2005, over twenty international conferences, workshops, and summer schools on networks took place; major meetings of electrical engineers, computer scientists, and physicists devoted many conference sessions to network discussions. At least four general audience books appeared, and the New York Hall of Science held a major exhibition titled "Connections: Seeing the World in a Different Way."[23] Major journals began to devote special issues, reviews, editorials, and covers to network science (*Nature Review Genetics*, *Nature Immunology*, *Nature Structural Biology*, and *Nature Reviews Systems Biology* all featured high-profile reviews, editorials, or covers; *Science* and *Proceedings of the National Academy of Sciences* both published special issues on networks).[24]

The interest in networks has not been limited to one field—network science is a major nexus of interdisciplinary work, with many of its key figures moving back and forth across applied mathematics, physics, computer science, engineering, economics, and sociology, as well as biology.[25] Network science involved the application of identical mathematical tools to cell phones, epidemics, computers, friends, proteins, and other sets of objects that could be imagined to be connected or associated in some way. By abstracting objects as points and associations between objects as connecting lines, such collections could all be depicted in the same manner and all became susceptible to the same tools. Network science allowed researchers (and laypersons) to imagine the burgeoning electronic networks and social and economic networks as the *same kinds of things*. The mavens of network science described the twenty-first century as a "connected age" in which computers, society, and biology all followed the rules of networks.

Studying the Internet became a way of studying networks more generally.[26] Like the telegraph in the nineteenth century, the Internet—as the preeminent communication technology of its age—became an "object to think with." That is, it became an exemplary object for analyzing all kinds of phenomena, including the dynamics of friendships, the economy, and politics.[27] The next sections provide several examples of how network science has been applied to biology.

Network Science in Biology

At the same time that network science began to gain interest and attention, some biologists began to announce moves away from the reductionist paradigms of molecular biology. In 2000, the first International Conference on

Systems Biology was held in Tokyo, and Institutes for Systems Biology were founded in that city and in Seattle, Washington. Systems biologists saw themselves as intellectual descendants of the cyberneticians Norbert Wiener and Bertalanffy, but they also traced their heritage to metabolic control theory and computer-mathematical simulations of populations and cells.[28] The new discipline promised a holist approach to biology that would integrate and make sense of the vast amount of data that had emerged from the Human Genome Project.

At first at least, there may have been little or no connection between network science and these moves toward holism. However, the reemergence of network thinking provided tools and opportunities for thinking about molecular biology and genomics in new ways. Network thinking suggested new approaches that biologists could take up as they struggled to make sense of genomes and genomic data. Networks provided an intellectual bridge between the Internet and biology; they played a crucial role in developing a "postgenomic" biology beginning in about 2000. The rise of the Internet did not *cause* a turn toward holism. But the Internet (via network science) did provide a ready-to-hand model (and set of tools) for biologists who were already searching for new approaches.

Network scientists were thinking about biology from the earliest days of their work. Watts and Strogatz's seminal paper had already shown that the *C. elegans* brain comprised a small-world network. It did not take long for Barabási to team up with the biologist Zoltán Oltvai in order to apply his scale-free network analysis to chains of metabolic reactions in organisms.[29] Barabási and Oltvai's paper is especially important because it can be read as an attempt to demonstrate the utility of applying network science to biology. The authors immediately invoke the "network" as the appropriate object of analysis. "Cellular constituents and reactions"—cells, mass, energy, information—are all "seamlessly integrated," the abstract tells us. The appropriate task is to understand their "large-scale structure," which is "essentially unknown."

Immediately, the paper invokes a new ontology: the components should be understood as part of a system. Understand the system, and it would be possible to see the role of the parts. In fact, even the networks themselves are not the right level of abstraction: because they are "extremely complex," they offer "only limited insight into the organizational principles of these systems." Just staring at the jumble of connections in the network itself is not going to tell you anything. Rather, Barabási and Oltvai's paper is about *measuring* the network, putting it in "quantitative terms" from which it becomes possible to understand its "generic properties." The network can only be understood as something that can be *reduced* to a particular set of mathematical quantities.

The main finding of Barabási and Oltvai's paper is a fairly straightforward one: the metabolic networks that they analyze turn out to be scale-free. That is, the biological networks seem to have the same structure that Barabási's team had discovered a year or so earlier. The biological networks have "striking similarities to the inherent organization of complex non-biological systems." In other words, metabolic networks look, from a particular point of view, rather like the World Wide Web, the Internet, and social networks. Such a finding, of course, justifies the enterprise of applying network science methods to biology in the first place. The power of the visual representations and mathematical tools of network science allowed scientists to begin to see similar networks everywhere.

Barabási and Oltvai also suggest that the scale-free-ness of the networks they have measured has a biological significance. It makes sense that biological networks are scale-free, they suggest, because this allows them to be "robust" and "error-tolerant." Biological networks, they reason, must have evolved over millions of years to be robust to environmental changes, and their analysis can detect this property. This conclusion depends on an analogy with the Internet: the ARPANET (which formed the original backbone of the Internet) was purportedly designed as a Cold War communication system, supposed to be robust to nuclear attack. Barabási and Oltvai's reasoning is that biological networks are similar to the Internet for a similar underlying reason—robustness.[30] Here, the demonstrated mathematical similarity between these different sorts of networks leads the authors to a finding about evolution.

But this conclusion relies on a form of circular reasoning: biological networks come to look like computer networks only when very specific features are picked out that make the two very different sets of objects look alike. This likeness is then put forward as evidence of common origin, justifying their similarity. Network science picks out very specific features of biological objects so that its mathematical tools can be applied. This abstraction requires biological networks to be represented in ways that are similar or identical to nonbiological networks. This similarity is then invoked to create and justify explanations and accounts of biological processes and action in nonbiological terms.

Another powerful approach based on network science—motif analysis—follows a similar pattern. In 2002, Uri Alon, a complex systems physicist at Weizmann Institute of Science, devised a detailed way of comparing biological and nonbiological networks. Alon and his coworkers thought that although networks might be similar on a large, statistical scale, their detailed structure might differ. In particular, they might have certain "motifs" that stood out—that is, particular small-scale patterns of connectivity. For instance, electronic circuits might have a lot of "feedback loops." Figure 6.2 shows all the possible three-node motifs (and one four-node motif). Alon and his colleagues devised an algorithm that could hunt around a network and count the occurrences of three-node and four-node motifs (more precisely, the algorithm hunted around the mathematical representation of the network stored as a set of nodes and edges in a computer). If a pattern appeared many times more than in a random network (used as a control), then it was considered to be a motif of that network.

Alon's study *did not* find much similarity between gene regulatory networks and electronic circuits or the World Wide Web. In gene regulatory networks, the most overrepresented motifs were "feedforward loops" and another motif they called the "bi-fan." But electronic circuits also exhibited three-node and four-node feedback loops, and the World Wide Web seemed to prefer "feedback with two mutual dyads," the "fully connected triad," and "uplinked mutual dyad" (these are just names given by Alon et al.; see figure 6.2). But Alon's conclusion was not that such network comparisons were useless. Rather, he suggested that the differences between the motifs represent the different functionality of these different networks. Indeed, the paper argued that motif analysis might be used to effectively characterize networks according to their function.[31] This reasoning suggests that if an electronic network could be found with motifs that matched a biological network, then it was likely that the electronic network and the gene network performed similar functions.

Although Alon found few similarities between motifs in the World Wide Web and motifs in biological networks, the paper reinforces the notion that electronic and biological networks are fundamentally similar. Electronic circuits, the World Wide Web, and biological networks are all "information processors," the paper argues, albeit ones designed for slightly different purposes. Indeed, the notion of "motifs" is premised on the idea that gene and protein networks are some kind of "logic circuits"—they can be understood in the cyber-electronic language of feedback and feedforward. The genome or proteome is made up of many of these overlapping, simple "motifs" that allow it to process information like a computer.

All these conclusions—both in Barabási and Oltvai's paper and in Alon's motif analysis—depend on two levels of abstraction. First, they depend on an abstraction of molecules, DNA, RNA, and proteins as mathematical points or nodes. The notion of a "motif" is once again a dramatic simplification

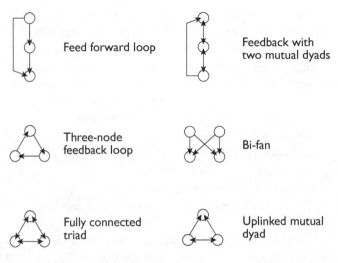

Figure 6.2. These specific patterns of connections were found to be overrepresented in various kinds of networks. Gene regulation and transcription networks exhibited feedforwards and bi-fans; electronic circuits exhibited feedforwards, bi-fans, and bi-parallels (not shown); and the World Wide Web exhibited feedback with two mutual dyads, fully connected triads, and uplinked mutual dyads. Credit: Adapted from Milo et al., "Network Motifs," table 1.

and abstraction: looking at the patterns of connections between molecules strips away any knowledge about the details of how the interaction occurs: for instance, the timescale or the strength of the interaction. It reduces all this to an edge (line) in a graph.[32] Second, network science depends on the reduction of networks themselves into quantitative and qualitative mathematical properties: size, density, average degree, diameter, clustering, robustness, and centrality, as well as notions such as feedforward and feedback.

Such analyses permit the comparison of vastly different kinds of objects (proteins, electronic networks, social networks) on the same terms. By describing biological systems in these highly abstract ways, it becomes possible to construct elaborate explanations and accounts of biology in distinctly *nonbiological terms*. By displaying the similarity of different sets of objects, the tools and representations of network science act to justify such accounts. Whereas before biologists might have seen distinct biological objects such as proteins or genes, network thinking allows them to perceive circuits, pathways, or groups; biologists could now describe and reason about biology in terms borrowed from electronics, computer networking, or social theory. The next section gives further examples of the kinds of accounts of biological action that such descriptions produce.

Wiring and Structure

Network science allows biologists to see biological systems in new ways: it emphasizes certain kinds of explanations and understandings of biological function and action. In particular, network thinking locates biological function in the *wiring* and *structure* of its parts. This is most often described in terms of "pathways." Here, function emerges not from any of the specific parts or even from collections of parts, but rather from the structure of their connections: the ways in which they are wired together.

Postgenomic biology has provided biologists with an array of new visualization tools that allow them to see and manipulate genomes (and other biological components) in different arrangements and configurations. The genome used to be linear—a one-dimensional text that could be mapped and sequenced end-to-end. This one-dimensional vision was central to the coherence of the genome projects, as well as to genetics throughout most of the twentieth century.[33] But the network invokes different kinds of spatial relationships—the networked genome is multiply connected in crisscross fashion. These connections are also dynamic, potentially changing over time. The proximity of genes or other sequence regions along the linear chain is less important than their proximity or "distance" between two parts in the network (indeed, "distance" has a special meaning in network science that has nothing to do with inches or nanometers: it counts the number of nodes along the shortest path between two points in a network).

One of the most iconic postgenomic representations of the genome is the circular visualization generated by Circos. This software tool displays genomic data in a ring, with lines crisscrossing the center showing various kinds of connections between different parts of the genome (figure 6.3). As its creators explain, "Data which represent connections between objects or between positions are very difficult to organize when the underlying layout is linear."[34] But such connections are precisely what postgenomic data sets consist of: alignments between different genomic regions, relationships between different genomes, genes implicated in the same disease, and interacting gene products. Circos images draw the viewer's eye toward the center of the ring, where the thickets of lines highlight the interconnectedness of the genome. The challenge such an image poses is one of trying to decipher the pattern or logic of these connections.

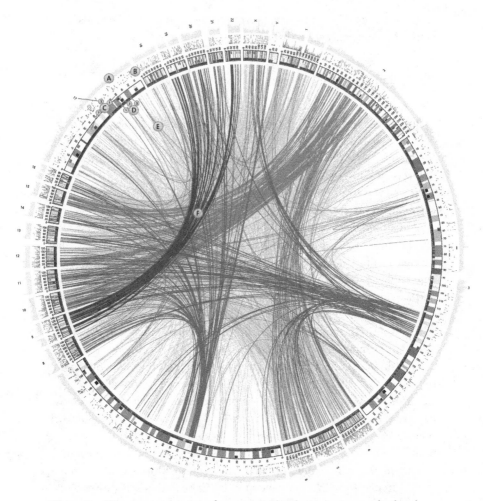

Figure 6.3. Circos representation of a genome. The chromosomes are displayed clockwise around the edge of the ring. The lines crossing the center of the ring link places on the genome associated with the same biochemical pathways. Credit: Martin Krzywinsky (http://circos.ca/). Krzywinski et al., "Circos."

Similarly, the methods of network science draw attention to the properties and structure of the connections between elements. Networks have allowed biologists to reimagine cells behaving as a society in which the individuals are cellular components. The cell-state metaphor is also an old one, having its origins in the nineteenth-century theories of Ernst Haeckel and Rudolph Virchow. For these biologists, the central thing was to be able to explain how cells could work together, cooperating to generate cohesive and functional multicellular organisms. For postgenomic biology, the challenge is to understand how the thousands of molecules in the cell function together to produce ordered and purposeful behavior. The Internet, the World Wide Web, and the social networks that exist on it provide the most powerful contemporary examples of such spontaneous order. The popular literature on network science emphasizes the importance of the highly connected behavior of large groups of people. Duncan Watts uses Napster (the now-defunct online music-sharing service) as an example: "The software that [Shawn] Fanning created was a neat trick, no doubt about it. But its enormous impact was a result not of any particular ingenuity of the code itself. . . . Rather, the magnitude of Napster's influence was a result of the huge numbers of people who realized that this was what they were looking for, and who started to use it."[35] The phenomenon arises spontaneously from the individual actions of all the people in the "crowd"; it is the structure of the links between them that generates the overall behavior. The idea of a network allows biologists to think about the functionality of cells and organisms in similar ways: it is not any individual part that is responsible for the overall behavior, but the way in which the behavior of the parts is *aggregated* by the structure that connects them.

Watts's work on "cascades" in networks provides a good example of how social dynamics are mobilized to understand biological systems. In 2002, Watts proposed a model for how information propagates through social networks. The aim was to understand the spread of rumors, the popularity of books and movies, and grassroots social movements. But "cascades" have also become a way of talking about biology: signals propagate through biological networks in the same way as fads propagate through social networks. This model has been used to try to understand how collections of proteins make "decisions" about responding to stimuli and how they maintain stability in noisy environments.[36] The overall effect is generated not by any individual element (just as one person cannot make a trend) but by the way in which the interconnections between elements allow signals to propagate.

Together, both the new visualizations and the new "sociality" of the genome generate new ways of organizing and using biological data from proteins, genes, and genomes. In particular, networks draw attention to complexity and combinatorics. In most cases it becomes clear that many genes, proteins, RNAs, or other molecules are involved in generating a phenotype. This might occur through cascades, or through other kinds of network interactions involving multiple elements. Using networks means becoming comfortable with such explanations, that is, with multiple elements acting together in ways that cannot always be completely unraveled, let alone predicted. The effect of "turning on" a gene becomes analogous to starting a rumor within a social group or posting a video on the Internet: the rumor may cascade through the network, or the video may "go viral," but it is hard to predict the cases in which this will occur. We need more information not about the gene (or the video) but about the structure of the network through which it propagates.

These network concepts have also been applied in the understanding of disease. In November 2000, the cancer researcher Bert Vogelstein published a *Nature* paper with David Lane and Arnold Levine arguing that cancer research too required more network thinking. Although the p53 tumor suppressor gene was first discovered in 1979, huge amounts of research had not resulted in a cure for cancer. Vogelstein and his coauthors suggested that the network was the problem: "One way to understand the p53 network is to compare it to the Internet. The cell, like the Internet, appears to be a 'scale-free network': a small subset of proteins is highly connected (linked) and controls the activity of a large number of other proteins, whereas most proteins interact with only a few others. . . . In such a network, performance is almost unchanged by random removal of nodes. But such systems contain an Achilles' heel: 'the most effective way of destroying a network is to attack its most connected nodes.'"[37] Curing cancer required understanding not only p53 but also the whole network of protein-protein interactions within which it was embedded. Reasoning about Internet hacking is made directly relevant to reasoning about biology: understand how to defend against cyber attack, and you will understand how to defend against cancer.

This kind of approach has been taken up in studies of breast cancer. In 2007, Trey Ideker's lab at the University of California, San Diego, examined whether network analysis could provide a better way of predicting which kinds of breast tumors would metastasize (finding reliable ways to predict metastases has important clinical implications). Usually, biologists have attempted to characterize tumors by examining gene expression profiles, that is, looking at which genes are turned on and off. Instead, Ideker's lab examined whether activities within "sub-networks" of genes could provide a better characterization of the tumors. Rather than looking for genes that were turned on or off, their method attempted to identify small networks (that is, groups of genes known to interact with one another) of genes that were turned on in metastatic tumors.

Their results showed that such small networks could be identified and that they did a more reliable job of predicting metastatic versus non-metastatic tumor types. "This supports the notion," they conclude, "that cancer is indeed a 'disease of pathways,' and that the keys for understanding at least some of these pathways are encoded in the protein network."[38] In cancer, the specific genes that are turned on may vary from patient to patient, but many of them are involved in the same pathways. This study is representative of network-based approaches in which specific genes, or even sets of genes, are no longer assumed to play the main role. Rather, it is the way in which genes (and proteins) are wired together that is taken to be determinative.

Network science has made such "pathways" visible and allowed network-based explanations of function to become plausible. Of course, individual genes and proteins remain at the root of these accounts, but the *kind* of causal explanation here is dramatically different, depending now on interactions between molecules and on notions of shape, structure, and pathway. What is new about these network approaches is not just their level of mathematical abstraction but also their connections to electronic and social networks. Linking biological networks to these other kinds of objects gives these kinds of structural- and wiring-based explanations greater plausibility and credibility.

Conclusions: Data, Complexity, and Reductionism

In biology, the network view enables a particular kind of reasoning—the abstract similarities between the World Wide Web, social networks, and biological networks suggest that we can learn about biology by knowing and thinking about the Internet and Facebook. Network science provides tools for thinking about how biology works: it works like an electronic circuit in a computer, or it works like the World Wide Web, or it works like a social network. And, in fact, these have all become the same thing—all manifestations of small-world information-processing structures. In other words, the most important achievement of network science in biology is to have created new sorts of objects: biological networks. These objects have been brought into existence through the representations and tools that network science offers.

But the usefulness and plausibility of networks in biology are also dependent on postgenomic technological infrastructures. Genomes are objects that are computationally reconstructed from thousands of (sequencing) experiments—we cannot "see" a genome except using the mathematical, statistical, software, and hardware tools that construct them in various ways inside computers.[39] Networks are similar: they are "empirical constructs" stitched together from hundreds or thousands of different experiments.

Data about protein-protein and protein-DNA interactions, for instance, are generated from assays such as a *yeast two-hybrid* system. In such tests, one pair of proteins can be tested to see whether they interact with one another. The genome of *Saccharomyces cerevisiae* (finished in 1996) allowed two research groups to attempt a complete mapping of the yeast protein-protein interaction network by running high-throughput yeast two-hybrid experiments (testing approximately $6,000 \times 6,000 = 36$ million potential interactions).[40] The resulting network, then, is the residue of such experiments, a way of summarizing this vast amount of data.

Complex networks are not very useful in and of themselves—graphs showing large numbers of interactions are often described as "hairballs," useless jumbles of connections. But network science provides a means of finding patterns in the vast amounts of data. Networks organize data. Other chapters in this volume and elsewhere show the importance of big data in postgenomic biology.[41] Next-generation sequencing machines and high-throughput experiments continue to generate vast amounts of biological data. Network science provides a set of techniques (and also a way of thinking) that synthesizes data and tries to make sense of it from a "thirty-thousand-foot" perspective. Network science is not just about drawing the map of connections between genes, proteins, and other molecules active in the cell. Rather, it is about using mathematical and statistical techniques to summarize and abstract away from these confusing webs.

In 2005, Evelyn Fox Keller suggested that the postgenomic welcoming of mathematicians and physicists—and their models—into biology had much to do with the growth of data and the "inability of conventional models to account for them."[42] The network is precisely such a "model"—it has allowed biologists to find new ways to analyze and synthesize and to explain biological function and action. More specifically, networks suggest the importance of *combinatoric* effects. The consequences of turning on a gene A may depend on the states of genes B, C, D, and E. A tumor suppressor gene, for instance, might be activated when A and B are on but C, D, and E are off, or when B and E are on and the rest are off. Again, it is the interconnections and structure of the relation between these elements that biologists increasingly see as crucial to understanding how the genome works and how diseases occur.[43] Networks make it easier to visualize how such systems might work; they provide a new way of thinking about how biological elements function together.

But knowing about all the interactions between genes A, B, C, D, and E requires lots of data. Network science not only might help biologists to process data and deal with complexity but also may actually help to encourage and justify the continuing data-greediness of biology. Acquiring more and more data only makes sense if there are ways to process and understand it. Network techniques are one of the tools that make use of these huge data sets, even relying on vast amounts of data in order to generate the statistical power that network science requires. Networks, then, are central to postgenomic biology not just because they provide an important set of tools, but also because they are bound up with the high-throughput, data-driven orientation of recent biological practices and knowledge.

Another way of expressing all this is to say that network science is a set of tools for dealing with complexity in biology. Large sets of genes, proteins, and other biological objects acting together constitute a complex system. Network science allows such systems to be abstracted in ways that not only help to organize data but also can help to build tractable mathematical models. The mathematical tools of network science are, in many ways, tools for analyzing complex systems—they construct explanations based on large numbers of objects, of potentially different kinds—acting together combinatorically (that is, nonlinearly).

Yet, this chapter has described in detail how network science is based on *reduction*—it abstracts biological parts into a few key elements and properties. Network thinking offers us complex, holist explanations of phenomena based on highly reduced models of biological objects. This apparent paradox may suggest that network science requires a new way of thinking about the relationship between reductionism and holism in biology. The key to resolving this dilemma is that network science is not in the business of reducing everything to genes or DNA: it is not necessarily *genetic* reductionism. Genetic reductionism omitted everything unrelated to genes (cells, cytoplasmic elements, environment, etc.) from its explanations.

Network science is engaged in different forms of reductionism—a reductionism to sets of abstract mathematical properties (nodes, edges, and their topological properties). This reductionism seems to be far more compatible with holist explanations—with accounts that involve large sets of heterogeneous parts and the relationships between them. At least since the middle of the twentieth century, reductionist and holist approaches in biology have stood in tension or opposition: the reductionism of molecular biology on the one hand versus the holism of biological systems on the other. But the introduction of network science in the past decade demands a rethinking of this binary. Systems biology also seems to have found some ways of reconciling reductionist and holist approaches.[44]

Rather than furthering the opposition between holism and reduction-ism, the reductionism of network science *enables* holist approaches and complex accounts. The information metaphor persists; understanding DNA and RNA remains at the center of much biological work. But overlaid onto this we have the notion that really understanding how all this informa-tion processing works involves understanding how organismic function is driven by networks, modules, and systems. Biologists must examine com-plex wholes, but in order to do so, they must model genes, proteins, and molecular interactions in simple mathematical terms. This modeling—and the abstraction on which it is based—also highlights the increasing inter-disciplinarity of postgenomic biology: biologists borrow and exchange tools with physicists, mathematicians, statisticians, and computer scientists. Net-work science has mediated the exchange of tools and ways of thinking (be-tween Internet studies and biology or between biology and social network analysis) that have become central to postgenomic biology.

These kinds of abstractions have become plausible not only because of linguistic or metaphorical connections between different kinds of net-works, but also through the kinds of common representations and tools that can be deployed. If a network can be an electronic network of comput-ers or a social network of people or a molecular network of proteins, then it must be only the abstract properties of these objects (computers, people, proteins) that are crucial for understanding the properties of the whole. Only when we simplify people and friendships to points and lines does the idea of a "social network" make sense. Likewise, simplifying molecular interactions to nodes and edges allows biologists to think holistically about organisms. Networks single out particular properties of biological objects for biologists' attention. Just as critiques of genetic determinism asked, "What is left out of informatic accounts of biology?" social scientists must now ask, "What is left out of network accounts of biology?"

Notes

1. In a protein-protein interaction, once the interaction is finished, there is no enduring link between the two molecules.

2. Some biologists and philosophers get around this by arguing that the nodes that constitute biological networks (proteins or genes) of this sort are "informa-tional." On this reading, A, B, and C are informational (rather than particularly physical) objects that can persistently interact with one another. In this case, bio-logical networks are understood as real networks of informational objects rather than metaphorical networks of real objects. For a discussion of this position see Godfrey-Smith, "Information in Biology."

3. On metaphor in science see Hesse, *Models and Analogies in Science*; Leatherdale, *Role of Analogy, Model, and Metaphor in Science*; Max Black, *Models and Metaphors*; Haraway, *Crystals, Fabrics, and Fields*; Keller, *Making Sense of Life*.

4. Dawkins, *Selfish Gene*, and *Blind Watchmaker*; Williams, *Adaptation and Natural Selection*.

5. Kay, *Who Wrote the Book of Life?*; and Keller, *Making Sense of Life*. Also see Lenoir, *Inscribing Science*.

6. Keller, *Making Sense of Life*, and *Century of the Gene*; Sapp, *Beyond the Gene*.

7. On the importance of visualizations see Daston and Galison, *Objectivity*, 2nd ed.; Wise, "Making Visible." On the epistemic importance of computers as tools in biology see Stevens, *Life out of Sequence*.

8. Keller, *Refiguring Life*, 108.

9. Otis, *Networking*. See also Lenoir, "Models and Instruments in the Development of Electrophysiology."

10. For more on cybernetics and its influence see Galison, "Ontology of the Enemy"; Pickering, *Cybernetic Brain*, and "Cybernetics and the Mangle"; Gerovich, *From Newspeak to Cyberspeak*; Heims, *Constructing a Social Science for Postwar America*.

11. Bertalanffy, "Outline for General Systems Theory," and *General Systems Theory*.

12. Holland, *Adaptation in Natural and Artificial Systems*.

13. Keller, *Making Sense of Life*.

14. For the most comprehensive account see Abbate, *Inventing the Internet*.

15. The city of Königsberg consisted of two islands in the middle of a river plus some of the surrounding mainland. The islands were connected to one another and to the mainland by seven bridges. Was it possible to walk through Königsberg crossing each of the bridges only once? (The answer was "no.") Biggs, Lloyd, and Wilson, *Graph Theory, 1736–1936*.

16. Erdős and Rényi, "On Random Graphs I."

17. Barabási, "Growth and Roughening of Nonequilibrium Interfaces,"

18. Keller has contested the claim that preferential attachment is the only way to achieve power-law distributions; Keller, "Revisiting 'Scale-Free' Networks."

19. The story is told in more detail in Keiger, "Looking for the Next Big Thing." See Albert, Jeong, and Barabási, "Diameter of the WWW"; and Barabási and Albert, "Emergence of Scaling in Random Networks."

20. Rapoport, "Mathematical Models of Social Interaction," and "Spread of Information through a Population with Socio-structural Bias"; Granovetter, "Strength of Weak Ties"; Milgram, "Small World Problem"; Simon, "Formal Theory of Interaction in Social Groups."

21. Rashevsky, *Mathematical Theory of Human Relations*.

22. A small-world network is one in which the number of nodes along the shortest path between any two nodes in the network is small relative to the size of the network. In social networks this phenomenon is sometimes described as "six degrees of separation." Watts and Strogatz, "Collective Dynamics of 'Small World' Networks."

23. Committee on Network Science for Future Army Applications, National Research Council, *Network Science*, 14–15.

24. Jasny and Ray, "Life and Art of Networks"; Shiffrin and Börner, "Mapping Knowledge Domains." Prominent reviews have also appeared in *Nature* (Strogatz, "Exploring Complex Networks"; Ottino, "Engineering Complex Systems"; Koonin, Wolf, and Karev, "Structure of the Protein Universe and Genome Evolution"), *Reviews of Modern Physics* (Albert and Barabási, "Statistical Mechanics of Complex Networks"), *Advances in Physics* (Dorogovtsev and Mendes, "Evolution of Networks"), *SIAM Review* (Newman, "Structure and Function of Complex Networks"), *IEEE Control Systems Magazine* (Amin, "Modeling and Control of Complex Interactive Networks"), and *Annual Review of Sociology* (Watts, "'New' Science of Networks").

25. See Committee on Network Science for Future Army Applications, National Research Council, *Network Science*, 14–15.

26. Typically, a 2010 textbook on networks used the Internet as its very first example: "One of the best known and most widely studied examples of a network is the Internet"; Newman, *Networks*, 3.

27. Including in several books aimed at a popular audience; see Watts, *Six Degrees*; Barabási, *Linked*; Buchanan, *Nexus*; Gladwell, *Tipping Point*; and Castells, *Internet Galaxy*.

28. Westerhoff and Palsson, "Evolution of Molecular Biology into Systems Biology."

29. Jeong et al., "Large-Scale Organization of Metabolic Networks."

30. Barabási elaborates on the connections between electronic, social, and biological networks in his 2004 popular science account (Barabási, *Linked*). For more details see Albert, Jeong, and Barabási, "Error and Attack Tolerance of Complex Networks."

31. Milo et al., "Network Motifs."

32. Uri Alon has since developed more sophisticated graphs and networks that take some of these elements into account; for a review see Alon, "Network Motifs."

33. See Rheinberger and Gaudilliere, *Classical Genetic Research and Its Legacy*.

34. See http://circos.ca/intro/circular_approach/. See also Krzywinksi et al., "Circos."

35. Watts, *Six Degrees*, 247.

36. See, e.g., Hooshangi, Thiberge, and Weiss, "Ultrasensitivity and Noise Propagation."

37. Vogelstein, Lane, and Levine, "Surfing the p53 Network."

38. Chuang et al., "Network-Based Classification of Breast Cancer Metastasis."

39. Barnes and Dupré, *Genomes and What to Make of Them*; and Stevens, *Life out of Sequence*.

40. Ito et al., "Comprehensive Two-Hybrid Analysis"; Uetz et al., "Comprehensive Analysis of Protein-Protein Interactions."

41. See in particular Leonelli, "Data-Driven Research in the Biological and Biomedical Sciences"; and chaps. 5 and 7 of this volume.

42. Keller, "Revisiting 'Scale-Free' Networks," 1067.

43. Maher, "Personal Genomes"; Tuch et al., "Evolution of Combinatorial Gene Regulation in Fungi"; Remenyi, Scholer, and Wilmanns, "Combinatorial Control of Gene Expression."

44. On systems biology see O'Malley and Dupré, "Fundamental Issues in Systems Biology"; Gatherer, "So What Do We Really Mean"; Wolkenhauer, "Systems Biology."

7

Valuing Data in Postgenomic Biology

HOW DATA DONATION AND CURATION PRACTICES CHALLENGE

THE SCIENTIFIC PUBLICATION SYSTEM

Rachel A. Ankeny and Sabina Leonelli

In the past fifty years, scientific publishing has undergone substantial changes as it tries to keep pace with the explosive growth of data and new technologies to share information in the digital age. This is particularly evident in experimental biology, where the genome sequencing projects of the late 1990s brought new attention and urgency to the challenges posed by large-scale data production and dissemination, as well as the relation between these activities and the publication of scientific papers. Online databases and repositories have gained recognition as important dissemination regimes for "omics" data, which complement the more traditional communication channel provided by scientific journals. In turn, this is affecting the very structure of journal publications, as well as the ways in which biologists negotiate and value authorship claims within them.

In 1953, Crick and Watson published their paper on the structure of DNA, which had just two authors, one figure, and no data and was only a page long. In 2001, the human genome was published in a paper that included 150 authors, forty-nine figures, and twenty-seven tables and was sixty-two pages long. In 2010, the 1000 Genomes Project was published as an open access paper available online and involved seventy-six institutions, had 12,145 SRA run IDs, and covered twelve pages. And in 2012, the ENCODE consortium published thirty papers from their efforts to describe all the functional elements in the human genome, each of which can be read as an "online package" complemented by extensive supplementary materials and sophisticated search mechanisms, so that users can select and retrieve information about topics of particular interest, such as "DNA methylation data."[1] These few examples alone suggest that the publication system, one of the most important mechanisms for the dissemination and rewarding of scientific achievements, is undergoing a major shift, which concerns not only the quantity of authors but also the ways in which authors are identified and publishable material is selected and presented.

In this chapter we reflect on the relation between this shift in authorship patterns and the ways in which research is conducted and assessed in contemporary biology. More specifically, we are interested in how the advent of large-scale, high-throughput data production (such as genome sequencing) and the dissemination of such data via online databases have challenged well-established norms for what counts as research outputs in biological research. At the same time as the reuse of data banked in large-scale databases is acknowledged to be central to the field's current practices, this trend remains in tension with existing intellectual and institutional structures, particularly credit attribution systems, which privilege authorship of scientific papers as the primary means of recognizing (and rewarding) scientific contributions and do not provide clear mechanisms for acknowledging data donation or curation.

Sociologists, historians, and philosophers of science have highlighted the key roles now being played by databases and other forms of electronic archives within the processes of scientific knowledge production and the rise of scientific fields devoted to the development and standardization of data dissemination practices.[2] These studies demonstrate that the use of databases is having a deep impact on how research is organized and carried out, yet this literature does not examine the implications of this widespread use of databases for current credit attribution and reward systems within science associated with authorship, particularly within molecular biology. We argue that paying attention to these scientific practices sheds light on one of the new and exciting aspects of postgenomic biology: the ways in which digital technologies are fostering new forms of labor and ideals of community and openness that differ considerably from the period that preceded the postgenomic era.

For our purposes, the term "postgenomics" signposts the most recent period in the history of the life sciences and begins at the point of the completion of the first genome sequencing projects in the late 1990s and early 2000s. As illustrated by a number of commentators, these projects did not constitute a fundamental rupture with previous research traditions in the life sciences.[3] However, the practices, conceptualizations, and technologies fostered by those projects have had a strong influence on how we understand biology today, particularly on the current emphasis on automated technologies for data production (such as next-generation sequencing machines), the key role of epigenetics and metabolic analysis in understanding organisms (which builds on and extends genomic results), and the social structures and mechanisms used to carry out research (which often mirror the centrally coordinated, yet multisited, division of labor used in international sequencing efforts).[4] Indeed, our use of the prefix "post-" is not meant to indicate that genomics is no longer part of biological research, nor that the techniques or methodologies developed in the previous "genomic" era have become unimportant. Rather, we use the notion of "postgenomics" as a historical marker for an era where the results of genomics, as well as the ways of doing and disseminating research that were developed in relation to the large-scale genome sequencing projects in the late 1990s, are being brought together with other types of biological traditions and outputs. Postgenomics in our view is thus primarily a signpost for a specific moment in the history of biology, whose distinctive characteristics can be analyzed in a variety of ways (as wonderfully demonstrated by the essays in this collection). Our analysis will be focused on the novel forms of labor and credit structures emerging from the central role taken by data practices, including both the production and the dissemination of data, within this period.

Data in Postgenomic Biology

The past decade has witnessed an unprecedented explosion of large-scale databases and related resources in the biological sciences, many of which are freely available via the Internet. Several scientists claim that their introduction heralds a new methodological paradigm in science, often referred to as data-intensive, or even data-driven, research.[5] Especially since the start of the Human Genome Project (HGP), data sharing, particularly in molecular biology, has become an almost unavoidable requirement, especially for those funded by public or governmental sources, participating in research consortia that require data sharing as a condition of membership, or submitting papers to journals that require data sharing as a condition for publication.[6] A variety of publicly accessible databases designed to house a range of types of molecular data arose in the late 1970s and 1980s, notably GenBank, which developed out of the Los Alamos Sequence Library (LASL), which was established in 1979.[7] Although some earlier databases existed, notably the Protein Data Bank, which has its roots in a paper atlas dating to the mid-1960s,[8] the rapid rise of molecular databases in this period can be traced to several factors, including the increase in the amount of molecular data being produced, advances in computer technology allowing for more efficient automation of both the production processes and the deposit procedures, and shifting norms about the extent to which researchers are expected to make their data publicly available. Also crucial was the widespread perception of online databases as a solution to the logistical challenges involved in multisited, cross-national collaborations, which are becoming increasingly popular as ways to structure and conduct research and require well-functioning infrastructures for the circulation of ideas, outputs, and resources.[9]

Key examples of contemporary databases that are well recognized both within and outside biology are those associated with genomic structure and sequence, such as GenBank, the EMBL-Bank database, WU Blast, and hundreds of others. Large-scale databases have become ubiquitous in biological fields ranging from ecology to evolution, and more broadly in the physical and chemical sciences. Combining data from more than one source using databases allows biologists to test hypotheses, obtain data to support empirical findings, and generate new research ideas more efficiently and on a broader scale than would be possible using more traditional methodologies.[10] It is widely recognized that unless data are appropriately handled, curated, and interpreted, the availability of large data sets could become an obstacle to knowledge integration rather than a contribution to its realization. The fact that data are produced by epistemically diverse groups, for different purposes, and in different parts of the world poses immense scientific, logistical, and structural problems when attempting to integrate these data.[11] Data dissemination via databases involves highlighting inconsistencies among sources—such as differences in the instruments used to produce data—through the adoption of standard terms and tools to capture data provenance. Other ways in which standardization occurs include elimination of differences in data formats and imposition of standards for what count as valid or reliable data. The process of integrating data within databases can thus be described as the gathering of different types of data, obtained by a variety of inconsistent sources, so that they can be searched and analyzed as a single body of information. Data donation and curation are an important part of this type of research activity, and it is therefore to those practices that we devote the bulk of our analysis.

Challenging What Counts as Research Output: Data Donation

The practices associated with data gathering and curation have changed significantly in recent times. Biological data curation arguably existed in various and less technological forms prior to the genomic era, and there are some points of continuity between contemporary practices used to build Internet-based databases and those utilized to construct earlier printed resources such as Margaret Dayhoff's *Atlas* or Victor McKusick's *Mendelian Inheritance in Man*.[12] In the earliest databases such as LASL, molecular data collection occurred based on traditional information science techniques of harvesting articles and abstracting from published journal literature. Staff working for the database searched the available scientific literature for sequence data and entered this information into the database, along with annotations detailing the citation information of the article in which the data were originally contained, the organism from which the data came, the location of the sequence within the genome where known, and other significant features of the sequence.

However, as genomic sequencing accelerated significantly, what became GenBank could not keep up with the pace at which data were being produced, and particularly with the annotation processes required.[13] Along with this shift also came a change in approach by scientific journals, which up until this period had been publishing sequence data as part of the evidence that accompanied research articles; they quickly came to recognize that once large-scale sequencing commenced particularly in the HGP, they would be unavailable to keep up with the resulting deluge of data.[14] So by the late 1980s, GenBank adopted an alternative model where scientists submitted their own data along with annotations directly to the database.[15] Other important databases such as the EMBL-Bank had similar structures.[16] Stephen Hilgartner convincingly argues that these two approaches (abstracting vs. direct submission of data) represent fundamentally different "communication regimes," a claim that we strongly endorse. As he notes, "GenBank could not simply wish direct submission into existence," but new relationships and incentives were necessary to induce participation in this new regime.[17] As we argue below, it was with these changes that the role of "data donors" as recognized today in molecular biology was formally created.

The role of "data donor" became codified not only through the change from abstracting to direct submission of data but also through a number of other changes that occurred in the 1990s. First, journals began to develop policies requiring authors to submit evidence that data relating to publications had been submitted as supplementary information and/or to appointed databases, or at least make explicit in their editorial requirements that the sharing of data (and other research-related resources) is an expectation, with a range of penalties for noncompliance.[18] In addition, a growing number of governmental grant funders put in place data deposit requirements, but often with allowances for researchers to retain data privately until publications based on the data appeared, with a typical maximum exclusivity period of one year (which later became six months).[19]

Both of these requirements were solidified by a voluntary agreement in 1996 among leaders of the HGP that all DNA sequence data should be released in publicly accessible databases within twenty-four hours after generation, a set of guidelines that became known as the "Bermuda Principles."[20] Some claim that the change to a twenty-four-hour rule fundamentally altered the landscape of data sharing in genomics,[21] as it cemented the values underlying the HGP (and which had their origins in model organism communities) of the project being a collective enterprise where participants recognize that scientific advancement depends on sharing sequencing data because this allows more efficient scientific practices and coordination of research efforts on a global scale. The Bermuda Principles seemed to codify an ethos that was more generally widespread with the rise of open source software and the "open science" movement, although there is limited evidence of any direct connections between these efforts.[22] It is undoubtedly the case that in the years following the promulgation of the Bermuda Principles, the immediate deposit of data into public repositories became a formal norm within genomics and in many other fields.

We thus wish to define data donation in the postgenomic era as the ensemble of activities through which a researcher submits a data set produced within his/her lab to an online database, with the goal of making that data set publicly and freely accessible. Since the early 2000s, these activities have been identified by most funding agencies within the United Kingdom, United States, and Europe as essential to the completion of a research project.[23] In other words, data donation is now viewed by public institutions as an important outcome of research in its own right: data are commodities that result from governmental investments in biological research, and as such making them publicly available is as important as publishing the key conclusions of such research in publicly accessible venues. What we want to stress here is that what makes data donation in the postgenomic era a novel phenomenon is not necessarily the specific practices involved in realizing it (though these have indeed shifted with time, to accommodate new technologies and venues). Indeed, they amount at least in part to a codification of practices already in operation within specific areas of biology, some of which were fairly formalized: for instance, consider the exchange of data, techniques, and strains of organisms enacted through newsletters and stock centers within model organism communities.[24] However, in contrast to the later structures and incentives associated with data donation, these traditions were in no way mandated and certainly were not enforceable: certain scientists and labs participated to greater or lesser degrees, depending on a range of other factors, and peer pressure was the main means of generating compliance. Most importantly, provision of data was part of what one did as a scientist who was a member of a broader scientific community, and was not tied into publication or funding mechanisms—in other words, data were not regarded as measurable research outputs in their own right. The introduction of direct submission at GenBank can thus be seen as a crucial step in a series of moves toward increasing the formalization and recognition of data donation as a distinct scientific activity, whose outputs count toward credit attribution and promotion criteria within academia.

One key motivation behind this shift in the status of data curation is the recognition that the activities involved are typically labor intensive and, as noted by Hilgartner back in his 1995 article, require relevant expertise. For a start, each database has specific guidelines on the standards, formats, and procedures to be used in order to submit a data set, and researchers wishing to contribute typically need to acquaint themselves with these guidelines and implement them. Further, preparing the data for submission requires annotating and formatting it, assessing which repository/database to use, and ensuring that submission of data complies with various requirements, notably immediate deposit rules. These are nontrivial tasks for the data donor. As automation in sequencing and in computing technologies has increased, these tasks have arguably become even more labor intensive, since donors need to identify appropriate standards for data production and formatting (or even develop such standards, where they are not already available) and train themselves to use these standards and include them in their daily research routine, so as to make the shift from data production to data donation as smooth and fast as possible. In some cases, such as the donation of a genome sequence to GenBank, the labor involved will be minimal, since the format of sequence data is highly standardized already and the parameters required by the database in order to be able to describe the data are minimal (indicating the species and strain of organism for each genome is typically enough for GenBank). In many other cases where more complex and less standardized data formats are involved, such as data about gene expression or development, much more labor is involved in donating data to a public database.

The high variability in the availability of appropriate standards is one of the key reasons why some types of data are easier to donate than others. Other factors include the variability of data themselves, the materials (specimens, tissue cultures, etc.) on which data are collected, the experimental setups used, and the instruments available to format data appropriately. For instance, if data are already produced in a digital format, as in the case of genome sequences, they are much easier to store in machine-readable databases than objects such as photographs of stains on an embryo. And indeed for transcriptomics data, which include various genomic and proteomic digital data sets, there is a global standard for annotation (Minimum Information About a Microarray Experiment [MIAME]); numerous internationally recognized databases for submission and storage, including Gene Expression Omnibus (GEO) and ArrayExpress; and a variety of tools for integrating and analyzing data, including the Nottingham Arabidopsis Stock Centre (NASC) Arrays and Genevestigator.[25] By contrast, metabolomics data are harder to standardize, and there is no agreed-upon way to annotate them, though there have been several initiatives, including the Metabolomics Society and ArMet. Cell biology images and phenotypic data are even more problematic: they have no generic or standard repository into which they can be deposited and do not have associated standards. There is also variation in how valuable specific data sets are considered to be. For instance, large data sets collected under standard conditions with good quality controls (such as the Arabidopsis transcriptomic data generated by the NASC in the United Kingdom) are viewed by both "wet" and "dry" researchers as key data sets for data-intensive approaches.[26]

Challenging What Counts as Research Activity: Data Curation

In addition to data donation, data curation also plays a crucial role in making data available and retrievable for analysis, particularly when carried out in the context of the construction and maintenance of online databases.[27] Database curators in particular serve as developers and long-term stewards for the databases used to disseminate data, which in turn are essential to the success of data-intensive research.[28] Examples of databases of key importance to contemporary postgenomic biology are "community databases" in model organism research, which bring together information about several aspects of a specific model organism (such as the Arabidopsis Information Resource, which collects sequence, metabolic, physiological, morphological, and expression data on the model plant *Arabidopsis thaliana*), and "grids" or "portals" in biomedical research (such as the Cancer Biomedical Informatics Grid, which purports to provide access to all sorts of data available on several types of cancer).[29] For data to become available through these databases, they need to be adequately prepared—in a more technical terminology, they need to be "curated" by professionals whose expertise lies in making those data accessible to further analysis.[30]

Curation today involves several labor-intensive tasks, including the selection of data to be assimilated into a database; their formatting into a machine-readable standard; their classification into categories linked to specific keywords, which makes it possible to search and retrieve data; and their ordering and visualization through "mining" tools that make it easier to spot meaningful patterns.[31] The activity of "data cleaning" alone, that is, the formatting and ranking necessary to prepare data for automated analysis, is estimated to take 80 percent of the total time invested in preparing data for mining.[32]

Consider, for example, the case of data that are obtained through in situ hybridization, an experimental technique used to establish whether a given gene regulates the development of specific morphological traits. The technique involves hybridizing a specific DNA molecule with an RNA probe, so that the mRNA produced by the DNA molecule will become stained and clearly visible under the microscope; gene expression, that is, the mRNA signals sent by DNA outside of the cell nucleus, is thus visualized. Where possible, this technique is applied to the whole embryo: researchers grow the embryo to a desired stage, inject it with the probe (figure 7.1), and, via a series of procedures designed to help the probe hybridize with the target DNA, are able to observe the parts of the embryo in which the chosen gene is expressed at different stages of development (figure 7.2). The resulting data come in the form of photographs of embryos at such developmental stages.

What we want to stress through this example are the complexities involved in curating data such as those resulting from in situ experiments. There are no clear guidelines or standards around what counts as a "good picture," and thus a valid data point, of an embryo, partly because this depends on the conditions under which the experiment is carried out, the ways in which embryos are kept, and how the gene expression stains manifest themselves. Curators in charge of integrating such data into an existing database need to exercise considerable skill in order to select data that they deem to be of high quality, as well as representative for a given biological object. When curators are dealing with data that have been already vali-

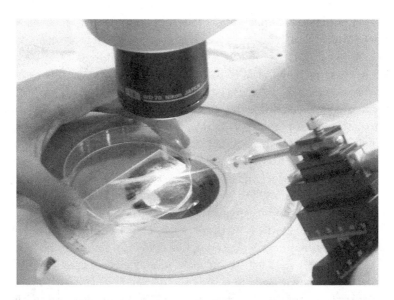

Figure 7.1. Researcher injecting a probe into embryos. Picture taken by Sabina Leonelli at the MRC Centre for Developmental Neurobiology, Kings College London (August 2007).

Figure 7.2. Each cylinder contains hybridized embryos at various stages of development. Picture taken by Sabina Leonelli at the MRC Centre for Developmental Neurobiology, Kings College London (August 2007).

dated as evidence for a publication, they can simply delegate responsibility for the quality of such data to the referees involved in that process; however in cases where data are donated prepublication, curators have a trickier job in selecting which data to include in the database, formatting those data in ways that make them accessible and legible to as many prospective users as possible, and labeling the data so that they can be easily found through an online search. A typical way to label data is what curators call a "unique identifier." This is a standard label given to the item with which the data set has been associated during data mining, so that there can be no confusion as to which item data sets are supposed to document. Importantly, the unique identifier needs to be machine readable, which makes it possible for data to be computed and analyzed through machines. The choice of unique identifiers marks the beginning of the process of annotation, whereby curators choose conceptual categories that (they hope) can be useful to order the database and make it intelligible to practicing biologists.[33] Again, how exactly to classify the relevance of data such as those obtained through in situ experiments is not straightforward, as it involves assessing the potential value of such data as evidence for a variety of biological investigations—a task arguably requiring in-depth knowledge of the research domains in question.

The complexity and conceptual effort involved into these activities illustrate the degree of responsibility that curators bear in setting up data for reuse, since it is clear that whoever chooses what counts as an adequate model, search mechanism, or visualization tool within a given database has a strong impact on how data retrieved through that database will be interpreted in the future. Indeed, experimental scientists are becoming increasingly aware of the extent to which curatorial decisions affect how data are eventually interpreted.[34] It has also been noted that the format given to data when they are processed for dissemination tends to drive the types of analysis that are then carried out—and thus the type of results obtained.[35]

Another challenging task for curators is selecting information to accompany data on their digital journeys. The interpretation of results obtained through databases depends heavily on the familiarity of researchers who use them with relevant experimental settings, instruments, and materials in actual laboratories.[36]

Professional curators are thus tasked with providing database users with information relevant to assessing the quality and reliability of data, which may include items such as the methods and protocols used to obtain data; the model organism, often down to the specific ecotype, used in the experiment; the instruments and techniques used; the publications or repository in which the data have first appeared; and the names and contact details of the researchers responsible for that bit of research, who can therefore be contacted directly for any question concerning information not directly reported in the database. Such information is often referred to as metadata, documenting the provenance of data.[37] Selecting and organizing relevant metadata is a task that, again, requires expertise in the scientific fields of interest, as well as significant conceptual choices in the classification and visualization of both data and metadata. When making these choices, curators come to play a crucial role in making it possible for experimenters to evaluate and interpret results obtained *in silico*, and set standards and criteria for what donors need to provide.

The above summary provides only a glimpse into the work carried out by data curators and its potential impact on how those data will be interpreted in the long term, and it suggests that their efforts will continue to be very difficult to fully automate. When selecting data formats, visualization tools, modeling techniques, and classification systems for data, curators are making choices that partly determine the significance that those data can have in the future. These choices are informed by the curators' own knowledge of scientific research and by their assessment of the potential usefulness of specific data sets toward new insights. Indeed, curators who are involved in developing particularly more complex biological databases, in which knowledge of the difference between data types and the significance of decisions made about how to visualize and integrate them is crucial, tend to have at least some training in experimental biology. Rarely are these curators trained only in computer science or information technology. Rather, it is often the case that they are lapsed biologists, who decided that curatorial work interested them more than laboratory work, or who thought that curation would represent a steadier and less time-consuming career than a career in laboratory science. The importance of curation work in biology and beyond is strongly underpinned by the emergence of training programs in data curation at leading institutions such as MIT and Harvard. Notably, these programs include a significant biological component, thus underscoring the significance of biological expertise in decision-making processes involved in data curation within this domain.

These reflections on the practice of data curation bring us to argue that the work done by curators can plausibly be viewed as an active contribution to the research efforts involved in interpreting the data, arguably on a par with the efforts carried out by researchers who access databases and compare the data found there with data they produce in their labs. When read in this way, data curation may be considered as an important contribution to processes of discovery in its own right, thus challenging existing assumptions about what counts as a knowledge-creating activity.

Credit Attribution for Data Donors and Curators

Within science, credit attribution is largely based on authorship of journal articles. Authorship is a formal acknowledgment of participation in and contribution to a particular research program and to a concrete product as represented in each publication. Credit attribution mechanisms shape the practices that govern hiring, promotion, tenure, and competitive grant processes, among other critical aspects of scientific practice and culture, and hence constitute a form of informal governance within the scientific community. Although there has been continued formalization of journal guidelines relating to authorship (e.g., the Uniform Requirements for Manuscripts Submitted to Biomedical Journals, or the "Vancouver Rules"),[38] the emphasis in these processes has tended to be on ethical responsibilities in publishing, including mechanisms for detecting and mitigating plagiarism, research fraud, ghostwriting, priority disputes, gift authorship, and disagreements over who is granted authorship. Little recent attention has been paid to norms for credit attribution specifically within molecular biology and the validity and usefulness of various credit attribution mechanisms,[39] despite the rapid cultural changes resulting from the growth of data-intensive practices and methods and their flow-on impacts. Researchers make important choices at all stages of research. Individuals are called upon to decide how to set up experiments and calibrate instruments that produce the data in the first place; how data should be formatted, mined, and visualized; and how data should be interpreted and which evidential value they acquire in different research contexts. Most of this labor is recognized in formal credit attribution mechanisms: coauthorship is defined by making significant conceptual contributions to both the research process and the writing up of its results (as indicated by the "Vancouver Rules"). Less formalized but still well recognized is the requirement to give credit in acknowledgments to technical support (e.g., for instrumentation assistance) or other types of labor.[40] This makes it all the more notable that activities such as data donation and curation, which, as we argued above, may well be seen as substantial contributions to the effective use of data as evidence for new discoveries, are not yet well acknowledged within existing credit systems. Indeed, there are few formal requirements in journal guidelines or elsewhere for the activities of data donors or curators to be either declared or acknowledged via any mechanisms, including authorship attributions. Neither type of contribution appears to consistently result in award of authorship on or even acknowledgment in peer-reviewed publications, and even where mechanisms are in place to credit data donors for their work, they tend to be inconsistent and highly variable in their application.

Credit attribution via authorship also involves an acknowledgment of responsibility and accountability, which is critical when attempting to establish a variety of norms relating to appropriate behaviors within the scientific research community. Lack of recognition of the labor involved in producing, curating, and disseminating data leaves data users without clear methods for establishing responsibility for data. This is highly problematic in cases where data may prove to be misleading or inconsistent, for instance, because of mistakes in curatorial practices or even intentional fraud. The labor associated with generating, curating, and disseminating data hence becomes a form of "invisible" labor.[41] In addition, lack of authorship attribution for those involved in generating, curating, and disseminating data also results in lack of endorsement of the appropriate (and inappropriate) ways of sharing and acknowledging data. Data are simply viewed as a resource to be used, without any explicit acknowledgment that they in fact have been produced through a complex series of processes performed by human actors.

The inconsistency of mechanisms for acknowledging and tracing data also generates problems in tracking reuse of data in the biological sciences and thus in assessing the scientific significance of the introduction of databases as key components of knowledge-production processes. It is extremely difficult to ascertain the impact of these databases on contemporary scientific practices, given that there are limited formal means by which feedback is given to curators about the usefulness of the databases and actual impacts of the data on scientific investigations and discovery.[42] More formal systems of credit attribution would permit more communication between data curators, donators, and users that in turn could improve systems for gathering, organizing, and sharing data. Hence, gaps in current credit attribution mechanisms have important epistemic implications beyond their impacts on incentive and reward structures: they shape the actual practices and results within the affected scientific fields, notably molecular biology.

In sum, data-based research strategies that are increasingly recognized as central to molecular biology, such as the consultation and analysis of large data sets retrieved from online databases, are in tension with existing intellectual and institutional structures relating to credit attributions. Although many grant funders require data deposit as a condition of funding, there are limited mechanisms for enforcing this, and no formal penalties for failure to upload and maintain data in public databases. In a scientific world that privileges authorship of journal articles as the primary means of recognizing and rewarding scientific contributions for a range of purposes, including tenure, promotion, and grant awards, the fact that data curation and donation are forms of invisible labor is highly problematic, as has been noted by scientists who themselves are involved in curation within fields closely related to genomics.[43] Value in science (be it of individual researchers or particular research projects) is largely calculated on the basis of the number of publications produced, the quality of the journals in which those publications appeared, and the impact of the publications as measured by citation indices and other measures: given that donation and curation are still largely unrecognized, the value of these activities correspondingly is limited in part because it cannot be measured using traditional metrics.

Why Are These Roles Not Considered to Be Part of "Authorship?"

As argued above, there is a clear absence of relevant institutionalized methods for credit attribution for practices associated with data. But what are the reasons for this gap? In this section we consider six possible answers to this question and conclude that some of these answers are much more plausible than others, particularly when we consider the continuities and discontinuities of contemporary postgenomic research with earlier phases of the history of biology.

We start by considering data themselves as the main protagonists of debates around what counts as contributions to current research practices. Can it be claimed that data in contemporary biology are fundamentally different (in some ontological sense) from other types of resources that historically were thought to constitute a contribution to research, such as strains, protocols, or unpublished results? It is unclear whether this argument holds: in many cases, provision of strains, protocols, or unpublished results was at least acknowledged in publication and in some cases was thought to constitute grounds for authorship (traditions, of course, differ from community to community). There are long-standing traditions (albeit inconsistently utilized or enforced) of acknowledgment for provision of "wet" resources (such as strains or samples) and of advice or mentoring, and authorship is common for those providing statistical or mathematical contributions to research. However, acknowledgment practices often have not extended to software and IT "support" for research. Perhaps more importantly, data have traditionally been published as evidence for a specific claim (if they are published at all), rather than as resources or outputs in their own right.[44] Hence, they do not fit conventional models of what counts as a scientific output.

A related argument often endorsed by science funders is that the scale—the sheer quantity of data—is different: the resources now available in the form of data are exponentially more extensive than what they were previously in resources that were more informally exchanged. But is this true, and what difference does this really make? It is certainly the case that new technologies for the production and storage of information are enabling scientists to retain a larger number of data than previously possible. However, while the importance of disseminating such data has been greatly emphasized, it is not clear that the ability to retain data has enhanced their value to scientists. It could also be argued that these data are now seen as less valuable, because they are being produced in part "just because they can be," often through automated means and not necessarily in a highly targeted manner as found in more traditional approaches to scientific work, such as using clear hypotheses or focused experimental procedures to determine what data to produce. These resources again come to be viewed as more technical than anything else, involving much less skill and art than other forms of data production, and in turn as less valuable. In addition, their ease of accessibility (in some cases, to anyone with an Internet connection) may also potentially decrease their value, which in turn decreases the value of the producers/donors and curators.

Another possible explanation is that data curation tends to be viewed as technical work, being done by nonprofessionals and in highly automated settings, and thus not as an integral part of scientific research. However, curators are often doctorate-level qualified biological scientists with experience of what end users need, serving in highly professionalized roles and doing "real" work under most definitions of what counts as science, much of which can hardly be relegated to machines. The scientific significance of this work does not necessarily match its visibility to the rest of the scientific community. Curation is often seen as a service or support activity that is distinct from research.[45] Indeed, there is a general lack of awareness among many data producers of the type of work that is involved in data curation. This is due in part to widespread ignorance, among experimentalists, of the skills and labor involved in data curation and the degree to which such skills and labor may ultimately affect data interpretation. Curation is more often viewed as background or routine work whose results are crucial to making data available but do not influence the ways in which data are analyzed and interpreted. Hence, data curation remains largely invisible to data users, who do not view the gathering and formatting of data as "real" scientific work involving conceptual decisions.

One further explanation frequently advanced with regard to the lack of credit attribution mechanisms for data donation and curation is connected to the requirement of data sharing that now is often part of public funding agreements. The activities associated with data donation are viewed by many as a burden and a waste of research time, which in turn leads to negative perceptions of the usefulness and value of databases. These issues again lead to data donation and curation being seen as side activities rather than the "core business" of science.

As Stephen Hilgartner and S. Brandt-Rauf argued in the early days of some of these debates, given the competitive nature of molecular genetics, there were some who wished to keep data private to maintain competitive advantage, let alone owing to concerns about commercialization, which are beyond the scope of this chapter.[46] As a result, GenBank's early rules allowed scientists to request that submitted data be kept confidential until the article based on that data appeared in print, and many may wish for a return to this regime. The switch to immediate banking of data on a daily basis, institutionalized through the Bermuda Principles and their influences on funders particularly from 1996 onward, and the idea that data should be shared insofar as they are "public commodities" resulting from "public investments" are very recent changes in scientific practice that may have contributed to a particular conceptualization of the types of activities associated with data donation and curation, a view of them as bureaucratic requirements rather than as a more informal and voluntary activity that nevertheless is an important form of scientific work.

This problem has been exacerbated by the historic absence of methods for measuring data reuse (some recent proposals that help to ameliorate this situation are considered in more detail below). Given that there appear to be large amounts of serendipity and "exploratory research" involved in utilizing existing databases to generate new results, patterns of use and reuse are difficult to ascertain. When data travel to research contexts beyond those in which they were originally produced, they are seldom "tagged" in consistent manners or attributed to particular researchers via authorship or other attribution mechanisms. Thus, it is difficult to determine which databases and curation methods are most useful or efficient for producing data in a manner that allows them to be reused, which in turn may well cause disquiet over the status and value of these contributions to research.

These issues point to broader tensions within the scientific community about various norms for the dissemination of scientific results and how they should be policed. Ongoing debates have focused on what researchers should be required to contribute back to the broader scientific community and importantly how such requirements should be enforced.[47] Despite what seems now to be widespread agreement about the need to share data rapidly and publicly, many researchers complain that enforcement mechanisms are weak or absent and that there are many ways to "game" the system. Others note that blanket agreements to deposit data fail to take account of the intricacies of commitments that researchers may have, for instance, to produce results from data that have been generated through private funds such as pharmaceutical companies, or even to others in their lab who are reliant on publications, such as early-career researchers. Recent shifts by governments and funding bodies to open access policies, such as the new policy enforcing open data for all research funded by U.K. research councils,[48] are increasing pressures in this domain, for instance, on publishers to ensure that the underlying data on which a scientific paper is based can be accessed by all. It clearly is well recognized that data access allows the scientific processes underlying research results to be scrutinized and reproduced, and research to proceed in a cumulative manner. Papers that have had data archived are not only more accessible to other scientists but are more frequently cited by them: for instance, a study found that papers that archived their data were cited 69 percent more often than papers that did not.[49] Yet only 25 percent of researchers currently make their data publicly available,[50] in some communities because of the scarcity of adequate repositories (less at issue in more genomics-related work), but also as a result of fear that their data will be misused, that errors will be found, or that they will lose their priority claims to the results that emerge out of the data. Hence, there remain divergent opinions on what behavioral norms are most appropriate and how they should be policed.

A key trend underlying this gap in credit attribution mechanisms is related to a more general shift in ideas of "gifting" within the postgenomic sciences. In the past, trading data (or protocols or other types of resources) was done based on a fairly simple reciprocity relationship between two investigators or perhaps two labs. However, donating data to a database for future use by an indeterminate and invisible group of other scientists, and for curating by people who are not viewed as part of the scientific community, clearly generates a much less personal set of contacts and in turn a different type of gift relationship. It is clear that the extent of personal connectedness differs according to database and community: in smaller-sized communities with databases focused on a particular organism (such as WormBase or TAIR), curators are often more active participants in scientific conferences and other research venues, generating a situation that is more akin to the traditional gifting relationship. But databases that are more heterogeneous in terms of their donors and users (such as large-scale biomedical databases) are based on a less clear set of preexisting communities and relationships and may reinforce the sense that data are no longer valuable, inasmuch as they are no longer being given as a gift by others doing similar research, nor is the gift relationship reciprocal, especially given the lack of enforcement mechanisms. In turn, then, authorship attribution or acknowledgement would not be viewed as appropriate, inasmuch as those doing the donation and curation work literally become invisible to the eventual user down the line, who in turn feels limited responsibility for acknowledging this research work.

A related issue is that these new forms of openness that we see in the postgenomic landscape and that arguably arose out of common norms in the small model organism communities in fact have started to erode the previous strengths of some of these communities. Once the sharing of information is not attached to specific communities with shared goals (and particularly in the case of more diverse uses of databases within biomedical research more broadly, including for commercial endeavors), it is, at least in principle, accessible to "all." Hence, donation and curation are not viewed as part of community activities, and the perceived distance between these activities and research makes it hard to consider data themselves as a form of research output.

Some experimental efforts are underway in other biological fields to attempt to mitigate some of the aspects of this situation, for example, via the Dryad Digital Repository. Dryad was developed by the National Evolutionary Synthesis Center and the University of North Carolina Metadata Research Center and is supported and financed by a number of journals and learned societies, particularly those in the areas of ecology, taxonomy, and evolution. To ensure that researchers are credited for reuse of the data deposited in Dryad, all data files are assigned a permanent, unique digital object identifier (DOI), which can be included in any future publications or with subsequent reuses of the data, and which allows access to the data. Researchers can archive their data within Dryad and be guaranteed of its future preservation, while satisfying various requirements from publishers and funders that they disseminate their research outputs.[51] Dryad also promotes adoption of its best-practice data citation policy and the traceability of data citations, provides a useful "safety net" for those data sets that do not have a home elsewhere, and prevents them from being lost when personnel leave a lab or a hard drive fails. However, there are concerns among researchers about whether it in fact contains enough metadata to allow in-depth reuse or reanalysis of data. The same strategy has been more recently adopted by FigShare, another system for the online dissemination of data that uses DOIs for citation, enables users to add a potentially infinite amount of metadata and accompanying information, and does not require data to have been published as part of a peer-reviewed journal article. The key issue for the purposes of this chapter is that Dryad, FigShare, and similar systems are introducing new forms of credit attribution that might present an alternative to authorship but currently do not generally result in the same sets of rewards that accrue based on authorship.

In addition to generic repositories, there has also been a recent emergence of "data journals," such as *Ecological Archives*, *ZooKeys*, *F1000Research*, *GigaScience*, *Database*, and *Earth System Science Data*. Publication in these journals is limited to data along with clear summaries of methods and metadata, with no conclusions being drawn from these data; hence, the focus of these new forms of publication is on increasing data accessibility, interpretability, and reuse. These journals therefore provide a mechanism for data producers and curators to be credited for their work by providing a means for attribution of effort for those who may not qualify for authorship on more traditional journal papers. Such journals also can help to encourage the concept that data are a research output in their own right, so long as reward mechanisms indeed recognize these publications as on par with more conventional forms of publication.

Conclusions: What Does Thinking about Credit Attribution Reveal about Postgenomic Science?

We have shown that research strategies arising in data-intensive biology that are widely acknowledged to be central to the field's current norms are in tension with existing intellectual and institutional structures, particularly credit attribution systems, which still privilege authorship of scientific papers as the primary means of recognizing (and rewarding) scientific contributions. Clearly, institutional structures lag behind: the publishing industry is notoriously conservative, as are academic reward systems and scientific funders. Also, there are important pragmatic obstacles to the facilitation and regulation of data dissemination; for instance, methods for tracking data use are as yet relatively primitive. More importantly, these are merely proxies correlated with deeper conceptual and epistemic trends within science. These trends include deeply held core ideas about the scientific method as being largely hypothesis driven rather than based on the accumulation of data and mining it for results.[52] These traditional ideals, in turn, involve a particular conception of what counts as a contribution to knowledge as relating primarily to discoveries, rather than, for instance, to underlying infrastructures such as databases, but also software and other specialized IT products that are essential to the practices of postgenomic biology.[53] Traditional ideas in science about conducting and replicating research also are in transition: data-driven research requires detailed documentation of provenance of data, as well as standardization of terminology, instrumentation, and so on, which in turn means that researchers need to know considerably more than just what each other's results are. Finally, there are persistent (and arguably antiquated) visions of what makes something scientific work (as opposed to technical work or support) that have been relatively unaffected by the rapid changes in technology, including what counts as a minimal publishable unit. Any moves to recognize curation and donation as activities that may require credit attribution in some form will simultaneously require the recognition of data curation and donation not just as forms of service activity but as forms of research in their own right.

What is remarkable about the current situation is that, for any given data set, numerous individuals, sometimes hundreds of them, are involved in making decisions about what to deposit, how to organize the data, and so on. Thanks to the unifying platform provided by computers and Internet access, those individuals are increasingly likely to have little in common: they probably will not know each other, they might have very different expertise and priorities, and they might be working within a variety of epistemic cultures. Most importantly, each of those individuals might have a very different set of tacit skills and embodied knowledge—and the ways in which they use data will reflect that diversity.

In the past, individuals pertaining to such disconnected communities would rarely have crossed paths. Thanks partly to the widespread use of digital databases, the division of labor within biology is becoming more fluid. The life of data is so long and unpredictable that there is no way to control who is manipulating them and how while they journey across the globe. Data users need to learn to interact productively with data curators in order to exchange ideas on how data could be visualized and experiments described, as well as provide feedback to visualizations, standards, and tools already in place. Mechanisms are in place to enable users to check for themselves the quality and reliability of data posted online—but while these tools are crucial to the interpretation of those data, users still need to trust producers and curators in their descriptions of their decision-making processes.[54] Many questions, of course, remain: for instance, what are the best mechanisms for recognition of donation and curation as an essential part of postgenomic research, and what should the metrics for quality and effectiveness be? Any attempts to institute new methods for credit attribution that take account of practices associated with data will require active debate and discussion about the nature and role of data in the postgenomic sciences.

Notes

1. See the ENCODE website, http://www.nature.com/encode/.
2. See, e.g., Hilgartner, "Biomolecular Databases"; Wouters and Schröeder, *Public Domain of Digital Research Data*; Bowker, *Memory Practices in the Sciences*; Hine, "Databases as Scientific Instruments"; Strasser, "Experimenter's Museum"; Stevens, "Coding Sequences," and "On the Means of Bio-production"; Leonelli and Ankeny, "Re-thinking Organisms"; Leonelli, "Packaging Data for Re-use," "When Humans Are the Exception," "Classificatory Theory in Data-Intensive Science," and "Making Sense of Data-Driven Research"; and Garcia-Sancho, *Biology, Computing, and the History of Molecular Sequencing*.
3. Barnes and Dupré, *Genomes and What to Make of Them*; Müller-Wille and Rheinberger, *Cultural History of Heredity*.
4. Hilgartner, "Constituting Large-Scale Biology: Building a Regime of Governance in the Early Years of the Human Genome Project"; Davies, Frow, and Leonelli, "Bigger, Faster, Better."
5. Kell and Oliver, "Here Is the Evidence, Now What Is the Hypothesis?"; Hey, Tansley, and Tolle, *Fourth Paradigm*.
6. Piwowar, "Who Shares? Who Doesn't?"
7. On this history, see Hilgartner, "Biomolecular Databases"; and November, *Biomedical Computing*.
8. See Strasser, "GenBank"; Stevens, "Coding Sequences."
9. Parker, Vermeulen, and Penders, *Collaboration in the New Life Sciences*.
10. Allen, "In silico veritas"; Wouters and Schröeder, *Public Domain of Digital Research Data*; Hey et al., *Fourth Paradigm*; Mariscal, Marban, and Fernandez, "Survey of Data Mining"; Tenenbaum, Sansone, and Haendel, "Sea of Standards for Omics Data."
11. O'Malley and Soyer, "Roles of Integration in Molecular Systems Biology"; Leonelli, "Integrating Data to Acquire New Knowledge."

12. See Strasser, "Collecting, Comparing, and Computing Sequences." See also McKusick, *Mendelian Inheritance in Man* and Its Online Version, OMIM," for a discussion of the evolution of this resource from a paper volume to online database. A detailed comparison of these resources and the technical, practical, and epistemological differences underlying their construction, maintenance, and use is beyond the scope of this chapter.

13. Lewin, "DNA Databases Are Swamped."

14. Hilgartner, "Biomolecular Databases."

15. Burks et al., "GenBank"; Cinkowky et al., "Electronic Data Publishing and GenBank."

16. For a general account, see Smith, "History of the Genetic Sequence Databases."

17. Hilgartner, "Biomolecular Databases," 253.

18. See McCain, "Mandating Sharing," for an extremely useful contemporaneous inventory and analysis of this issue.

19. See National Research Council, *Sharing Publication-Related Data and Materials,* and *Ensuring the Integrity, Accessibility, and Stewardship.*

20. "Summary of Principles Agreed upon at the First International Strategy Meeting on Human Genome Sequencing"; Contreras, "Bermuda's Legacy"; K. M. Maxson, R. A. Ankeny, and R. M. Cook-Deegan, manuscript in preparation.

21. See, e.g., Contreras, "Bermuda's Legacy."

22. K. M. Maxson, R. A. Ankeny, and R. M. Cook-Deegan, manuscript in preparation.

23. Leonelli, "Why the Current Insistence on Open Access to Scientific Data?"

24. McCain, "Communication, Competition, and Secrecy"; Star and Ruhleder, "Steps towards an Ecology of Infrastructure"; Ankeny and Leonelli, "What's So Special about Model Organisms?"; Kelty, "This Is Not an Article."

25. Websites for these tools and databases are available, as of January 6, 2014, at the following URLS: http://www.mged.org/Workgroups/MIAME/miame.html (MIAME), http://www.ncbi.nlm.nih.gov/geo/ (GEO), http://www.ebi.ac.uk/arrayex press/ (ArrayExpress), http://arabidopsis.info (NASC), https://www.genevestigator .com/gv/ (Genevestigator).

26. Leonelli et al., "Making Open Data Work in Plant Science."

27. Blake and Bult, "Beyond the Data Deluge"; Stein, "Towards a Cyberinfrastructure for the Biological Sciences"; Renear and Palmer, "Strategic Reading, Ontologies, and the Future of Scientific Publishing."

28. Howe et al., "Big Data."

29. Huala et al., "Arabidopsis Information Resource (TAIR)"; von Eschenbach and Buetow, "Cancer Informatics Vision."

30. Blake and Bult, "Beyond the Data Deluge"; Buetow, "Cyberinfrastructure."

31. Howe et al., "Big Data"; Leonelli, "Packaging Data for Re-use."

32. Börner, *Atlas of Science.*

33. One of us has discussed the problems, opportunities, and tensions associated with this process in great detail, with specific reference to the case of bio-ontologies, a classification system used to order data in many biological databases;

see Leonelli, "Centralising Labels to Distribute Data," and "Classificatory Theory in Data-Intensive Science."

34. Mariscal et al., "Survey of Data Mining."

35. Fry, *Visualising Data*.

36. Leonelli, "Packaging Data for Re-use," and "When Humans Are the Exception."

37. Edwards et al., "Science Friction."

38. International Committee of Medical Journal Editors (ICMJE), "Uniform Requirements for Manuscripts Submitted to Biomedical Journals."

39. McCain, "Communication, Competition, and Secrecy," examined related issues over twenty years ago.

40. Cronin, *Scholar's Courtesy*.

41. Shapin, "Invisible Technician."

42. Bastow and Leonelli, "Sustainable Digital Infrastructure."

43. See, e.g., McDade et al., "Biology Needs a Modern Assessment System."

44. Leonelli, "Packaging Data for Re-use."

45. Leonelli, "Packaging Data for Re-use," and "When Humans Are the Exception."

46. Hilgartner and Brandt-Rauf, "Data Access, Ownership, and Control."

47. Longino, *Science as Social Knowledge*; Radder, *Commodification of Academic Research*.

48. See policy document, available as of May 23, 2013, at http://www.rcuk.ac.uk/RCUK-prod/assets/documents/documents/RCUKOpenAccessPolicy.pdf.

49. Piwowar, Day, and Fridsma, "Sharing Detailed Research Data."

50. See http://www.parse-insight.eu/project.php.

51. Whitlock et al., "Data Archiving."

52. O'Malley et al., "Philosophies of Funding"; Leonelli, "Classificatory Theory in Data-Intensive Science," and "Making Sense of Data-Driven Research"; Callebaut, "Scientific Perspectivism."

53. McDade et al., "Biology Needs a Modern Assessment System."

54. Leonelli, "Packaging Data for Re-use," and "When Humans Are the Exception."

8

From Behavior Genetics to Postgenomics

Aaron Panofsky

It is no exaggeration to label behavior genetics the quintessentially controversial science. Behavior geneticists have never fully escaped the long shadows that eugenics and scientific racism have cast over their subject. Efforts to identify the "genetic underpinnings" of culturally sensitive topics such as intelligence, criminality, sexuality, mental health, and personality have always raised accusations of genetic determinism: if traits or differences have genetic causes, this suggests that inequality is natural and amelioration might be impossible or even dysfunctional. The science of behavior genetics has been subjected to intense scrutiny and criticism. Critics have charged a radical mismatch between behavior geneticists' ambitions and what their science will warrant.

The core tool of human behavior genetics has been and continues to be family studies, which are used to partition the variation of a trait observed in a population into genetic and environmental components. By comparing the correlations of IQ scores, say, among monozygotic and dizygotic twins, other siblings, and parents or among adoptees, birth parents, and adopted family members (and making a series of often-controversial assumptions), behavior geneticists can estimate heritability, or the proportion of trait variance due to genetic variance. These methods have proven incredibly fruitful, yet they have opened behavior geneticists up to attack. Of the many charges leveled, one set is especially relevant for our purposes: these methods infer "the genetic" indirectly through family relationships, never measuring DNA or biological material of any kind. Heritability estimates do not help identify particular genes or ascertain their functions in development or physiology, and thus, by this way of thinking, they yield no causal information.

The advent of new molecular genetic technologies in the late 1980s and their rapid subsequent development seemed to make feasible the direct linking of DNA to human behaviors. These began with linkage studies that used family pedigrees and genetic markers to link behavioral traits and disorders to chromosomal regions. Researchers have searched for behavioral quantitative trait loci (QTLs) by comparing cases and controls in a variety of association studies, from probes of individual candidate genes all the way to genome-wide association study (GWAS) scans of hundreds of thousands of single-nucleotide polymorphisms. Since the late 2000s, researchers have begun investigating behaviors, especially pathologies, with next-generation sequencing techniques looking for rare variants, microsatellites, and de novo mutations. So too are they seeking to use techniques often labeled "postgenomic"—searches for genetic signals of selection, functional genomics, and epigenetics. As they have become available, scientists have consistently applied new molecular techniques to behavior.[1]

Would the dramatic technological advances of the molecular and genomic era be able to pull behavior genetics from the quagmire of controversy? There were three basic ideas about how this might occur. First, even staunch critics of behavior genetics, such as geneticist Richard Lewontin, had identified the capacity to measure DNA as crucial to judging the field a legitimate science.[2] Second, a related hope was that molecular research could unify the epistemologically fragmented field. Unlike heritability, the basic interpretation on which scientists could not agree, DNA, with its seemingly unambiguous material nature, might provide a common framework for research. A third hope was that molecular genetics would upgrade the field's professional profile. Few human behavior geneticists to that point were disciplinarily identified with biology or genetics. Most were psychologists, and critics sometimes hinted they were out of their scientific depth. Molecularization could improve behavior genetics by bringing in researchers with rigorous biological expertise and by enabling integration with biological disciplines that had never been achieved. Behavior geneticists have thus eagerly heralded molecular research as a "new era" that will "revolutionize" the field.[3]

On the other hand, some have viewed the molecularization of behavior with trepidation.[4] They worry that molecular genetics would reignite claims of the genetic determinism of behavior. If things were bad when we were talking metaphorically about genetics via twins, imagine the potential for mischief when actual segments of DNA are involved. DNA correlated with behavior is still tricky to interpret. The association could be spurious, an artifact of the ways subjects are identified or behaviors are measured. Simply identifying a gene or a chromosomal region does not say what the gene does or how it might influence behavior. Genes always act in concert with many factors and usually have only a probabilistic influence on an outcome. And, as the example of phenylketonuria (PKU)—a genetic metabolic disease treated by a dietary regime—shows, even a highly genetic trait might be changed by an environmental intervention. Critics feared that this highly technical, yet glamorous, science would be easy for journalists and the public to misinterpret, and that as molecular genetics got more attention and became more costly and scientific competition increased, scientists would have more incentive to make bold, provocative announcements.

This chapter shows that neither the hopes nor the fears were realized. Molecularization—from linkage and association studies to GWAS and other postgenomic techniques—failed to quell the field's epistemic problems or quiet its controversies. But neither did it harden behavior geneticists' tendencies toward genetic determinism. Quite the opposite occurred. The older quantitative genetic claims of behavior genetics involving twins and family studies became a relative oasis of certainty and consensus as molecular genetics produced ambiguous results. And many behavior geneticists became more reluctant to make genetic reductionist claims about behavior and became much more interested in looking at *environmental* causes of behavior. Why?

There are two intertwined reasons for this counterintuitive outcome. The first is the dramatic failure of molecular technologies to fulfill anything close to the expectations that had been put on them. Molecular methods have proven limited at finding genetic correlates of behavior and terrible at meeting standards of replication. There have been similar problems across genetics, but behavioral traits, which are more difficult to define than heart disease or diabetes, have been particularly vexing. One might think that the technological quandary is sufficient explanation; after all, the problematic methods cannot quell controversy, nor are they a sound basis for making deterministic claims. But a crucial factor was the field's social structure and competitive dynamics.

When the geneticists and psychiatric researchers bearing molecular methods entered behavior genetics, they encountered a disciplinarily fragmented field. Rather than joining the preexisting behavior geneticists or working to build serious social connections, they remained sequestered in their own scientific communities. Though there was a great deal of consensus about the value of molecular methods and especially how to cope with their disappointing results, this did not overcome the competitive dynamics among different groups. The arrival of molecular genetics presented the veteran behavior geneticists with a dilemma. The technology presented an obvious opportunity to explore new research questions and collaborations. Yet competition with the new contingent of molecular genetic hotshots also threatened to devalue their expertise and scientific investments. Thus, in addition to taking on molecular genetics, they also sought to distinguish themselves. They did this by reimagining themselves and their research in environmentalist terms while subtly casting molecular geneticists as crude reductionists.

In this chapter I eschew much of the complexity and heterogeneity in behavior genetics by focusing on two groups' reactions to the advent of molecular technologies: veteran behavior geneticists (mostly psychologists who had been studying humans using quantitative genetic techniques) and newcomers from psychiatry and human genetics. Veterans and newcomers shared an enthusiasm for molecular techniques and eagerly sought to apply them to the gene/behavior riddle. But disciplinary differences and the field's legacy of controversy made them mutually suspicious as well. As molecular techniques disappointed expectations, veterans and newcomers (who rarely embraced the "behavior genetics" label) reacted differently. The fact is, behavior genetics is a deeply heterogeneous and contested space where many different kinds of scientists compete to define good science but often struggle to avoid becoming explicitly associated with it.[5] It should be noted that here I am using the term "behavior genetics" largely to refer to those quantitative genetics-wielding psychologists who had self-consciously claimed the label rather than anyone whose research might substantively address the gene/behavior question. Thus, in the service of the narrative, I smooth over some of the complexity and heterogeneity among those engaging the genetics of behavior.

High Hopes for Molecular Genetics

The late 1980s and early 1990s were an exciting time in genetics. The invention of new techniques and technologies for manipulating DNA were opening up tremendous possibilities for research. The ambition of geneticists was to know everything about the human genome. But such an effort would be costly in time, money, and expertise. Genetics leaders mobilized the promise of unraveling the causes of behavior to convince the public that it would be worth spending $3 billion over fifteen years to conduct the Human Genome Project (HGP).[6]

The genetics profession had been ambivalent about studying human behavior since distancing itself from the eugenics movement in the 1940s.[7] Field leaders swallowed these compunctions, however, and vigorously promoted unlocking the secrets of human nature—that is, behavior—as the raison d'etre for the HGP. In his humbly titled essay, "A Vision of the Grail," the eminent geneticist Walter Gilbert argued that the HGP will answer the question, "What makes us human?"[8] James Watson, codiscoverer of DNA's structure and first director of the HGP, often proclaimed, "We used to think our fate was in the stars, now we know it to be in our genes."[9] The "horrors of the deranged mind," he also stated, "more than give us reason to find the genes we know are there."[10] Biochemist Daniel Koshland, editor of *Science*, frequently used his editorial page to champion the HGP and to argue that its impact would be tremendous on crime, homelessness, and other social problems related to mental illness. Social welfare and public safety programs, he claimed, are "Band-Aid" solutions that fail to address the real, genetic roots of the problem.[11]

For many geneticists the appeal of studying behavior was undeniable. Accustomed to working on publicly inscrutable molecular mechanisms, behavior genetics offered them the opportunity to attract public and media attention and present themselves as authorities on "what makes people tick," as Dean Hamer, father of the "gay gene," put it.[12] The words of Watson, Koshland, and others were more than a sales pitch; they were a road map for geneticists to insert themselves at the center of crucial public issues, with therapeutics, not eugenics, being the aim.

This burgeoning of interest in behavior genetics brought a rush of resources and personnel to the field. Research funding poured into the field. Figure 8.1 shows the number of grants made by the National Institutes of Health (NIH) listing "behavioral genetics" as a keyword. From the mid-1970s to the mid-1980s, there was a modest, but declining, number of grants being funded. From 1985 to 1990, only a handful of grants were made; however, starting in 1991, the era of molecular genetics, NIH grant making rapidly increased.[13] The cumulative amount of money spent on new grants for behavior genetics from 1997 to 2005 was over $600 million for 558 separate grants to 491 different principal investigators.[14]

This expansion in funding support was accompanied by an expansion of scientific institutions as well. In 1992 psychiatrists interested in genetic research founded the International Society for Psychiatric Genetics, which

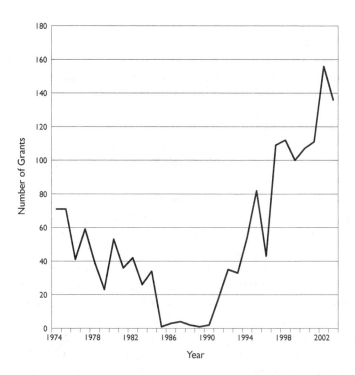

Figure 8.1. NIH research grants in "behavioral genetics" (1974–2003).

sponsored the *American Journal of Medical Genetics Part B: Neuropsychiatric Genetics*. In 1996 a group of animal behavior geneticists and genetically oriented neuroscientists formed the International Behavioural and Neural Genetics Society, and in 2001 it launched its own journal, *Genes, Brain, and Behavior*.[15] Several other journals were founded around this time, including *Psychiatric Genetics* (1990), *Molecular Psychiatry* (1996), and *Neurogenetics* (1997). These joined the Behavior Genetics Association (BGA) and *Behavior Genetics*, which had represented the field since 1970.

This growth brought many new scientists to the field, though precise numbers are unavailable. In 1985 the BGA had 388 members, and this swelled to 578 in 1996.[16] The combined membership of the three main societies in the late 1990s was probably between one thousand and fifteen hundred.[17] The nearly five hundred principal investigators of funded grants offer another estimate since each is likely to lead a group of several other researchers (junior colleagues, technicians, postdocs, and graduate students); perhaps fifteen hundred to two thousand researchers were working in behavior genetics from the late 1990s to early 2000s. Thus, though behavior genetics had grown rapidly during this period, it was still a small-to medium-sized field.

The new entrants to the field had different disciplinary backgrounds and expertise—more medical and molecular—than the long-standing behavior geneticists. While this certainly created tensions, field veterans were able to access the new research tools. Some certainly retrained, but most were able to collaborate with molecular genetics experts. In many ways the "molecular revolution" in genetics was not conceptual but technological. Many experimental processes were black-boxed and automated, that is, they could be deployed without their technical complexity or ambiguities being engaged. Less and less was meaningful work necessarily tied to "wet lab" expertise, which could often be outsourced.

As I have shown elsewhere, motivations varied for combining molecular genetics with behavioral topics.[18] Veteran psychological behavior geneticists were eager to pursue new scientific opportunities with molecular methods, but they were also wary of being sidelined by the incoming geneticists. Psychiatrists were eager to pursue genetics and envious of the seeming progress of genetics in other medical fields. As the psychiatric geneticist Kenneth Kendler has reflected, "[There was] rapid, and nearly relentless, success of human genetics in mapping the classic Mendelian human genetic disorders. Few in the psychiatric genetics community could avoid feeling envious as these major disorders were mapped one by one. There was a great desire to get into the line with the hope that schizophrenia, manic-depressive illness, panic disorder, or alcoholism would be next."[19] Could genetics push the behavioral and psychiatric sciences into the limelight? Molecular geneticists acted, with more than a hint of arrogance, as if they could swoop in and crack problems that had long bedeviled behavior genetics. Thus, molecular genetics brought incredible enthusiasm to behavior genetics, but it also spawned competition and tension among those acting in this scientific space.

Behavior genetics was not a very "fielded" field.[20] This is to say that there were few barriers to entry, there was little mutual policing, and scientists tended to see the point of doing behavior genetics as competing for recognition in other scientific fields rather than the recognition of other self-identified behavior geneticists. The new scientists coming into the field tended not to identify or integrate with the veteran behavior geneticists. And though their scientific activities affected each other—a finding on one side of the field influenced the plausibility of what those on the other side were doing—there were few mechanisms of accountability to each other.

Failures of Molecular Genetics of Behavior

Molecular genetics has been a major disappointment, if not an outright failure, in behavior genetics. Scientists have made many bold claims about genes for behavioral traits or mental disorders, only to later retract them or to have them not replicated by other scientists. Further, the findings that have been confirmed, or not yet proved false, have been few, far between, and small in magnitude. Molecular studies of behavior have often been inspired, on the one side, by universally high estimates of heritability for behaviors and, on the other, by the identification of genetic mutations for certain forms of mental retardation (like Fragile X and PKU). But there has been a tremendous mismatch between these successes and the limited achievements of molecular studies of behavior.

It is important to note that the disappointments or failures of the field are by the practitioners' own standards. There have been *additional* critiques, for example, of the decontextualized definition of behaviors that are historically and culturally specific. Critics have also disputed the usefulness of searching out genes without a model of the organism's development.[21] But here I will consider prominent examples that have been failures in the molecular researchers' own terms: the replicability of claims and the statistical significance and magnitude of effects.

An early finding heralding the promise of the molecular era was Janice Egeland's 1987 announcement of a genetic linkage to manic depression found by analyzing the pedigrees of Amish families.[22] But just two years later the finding was retracted when a member of the family from a different branch of the family tree was diagnosed, which rendered statistically insignificant the link between the putative chromosome location and disease in the family line.[23] When the finding was reported in a conference, the eminent molecular geneticist David Baltimore remarked, "As an average reader of *Nature*, what am I to believe?"[24] It would get worse: reviewing nineteen other failed bipolar disorder linkage claims in 1996, geneticists Neil Risch and David Botstein wrote of the "euphoria of linkage findings being replaced by the dysphoria of non-replication . . . creating a roller coaster-type existence" for field members and observers.[25] Molecular genetics research on manic depression remains unsettled today.[26]

In 1993, cancer geneticist Dean Hamer's group announced, to tremendous public interest (both fascinated and outraged), a link between a region on the X chromosome and male sexual orientation in the population he studied.[27] Academics and the lay public heatedly debated the scientific and social meaning of the claim.[28] In an unusual move, Hamer cowrote a popular book, *The Science of Desire*, detailing his process of discovery.[29] But another lab's attempt to replicate the finding on an independent sample failed.[30] Scientifically, this claim has largely faded away, though it lives on in the public imagination.

Another focal point for molecular genetic research has been alcoholism and alcohol use. Much early attention was focused on research by Ernest Noble and Kenneth Blum, who claimed to find an association between alcoholism and variants of genes that produce a receptor for the neurotransmitter dopamine.[31] This suggested not only a genetic link but also a pathway for uncovering a molecular mechanism involving the brain's ability to produce "pleasure." Journalist Constance Holden reported the history of failed replications and even the growing annoyance of Noble and Blum's colleagues that they continued to forcefully press their case.[32] Sociologists Peter Conrad and Dana Weinberg put this research in a broader context, arguing that media coverage had given the impression that the "alcoholism gene" had been discovered three times by 1996.[33] Further, a small set of researchers had trumpeted their supposed success repeatedly with little reference to the unsettled state of knowledge.

One of the most celebrated recent behavior genetic studies, about a gene-environment interaction in depression, was led by Avshalom Caspi and Terrie Moffitt.[34] The researchers found that differences in the promoter region of the serotonin transporter gene mediate the effect of stressful life events on depression. People who experience stressful life events are much more likely to become depressed if they have one or two copies of the "short" allele compared to those with two copies of the "long" allele. *Science* mentioned it among the most important scientific discoveries of 2003.[35] However, attempts to replicate the study have been mixed. In 2009 Kathleen Merikangas and Neil Risch pooled the data from fourteen studies into a meta-analysis that found no effect of the gene regardless of life experience.[36] Epidemiologist Stanley Zammit and colleagues have looked at the recent trend of studying gene-environment interactions (as in the Caspi study) in behavioral traits. They argue that claims of replication have not been warranted and, indeed, that the whole endeavor is of limited value to understanding the cause and treatment of mental disease.[37] This promising line of research, widely attractive for focusing on interaction rather than genetic determination, has also been derailed.

These examples are part of a broader pattern of disappointing results. In 2002, geneticist Joel Hirschhorn and his colleagues performed an exhaustive review of genetic association studies. Of the 600+ reported associations to any medical trait (20 were psychiatric), 166 had been studied three or more times and only 6 had been consistently replicated.[38] This was an early entry in what has become a parallel literature reporting molecular genetic disappointments and offering instructions for cautious interpretation and improvement of the corpus.[39] By this point a secondary backlash literature has emerged, charging the field with being now too stringent and sowing doubt with "pseudo-replications."[40]

These disappointments have also afflicted postgenomic behavioral studies. GWAS have evolved beyond scans seeking to find one or a few genes associated with a trait toward analyses of networks of interconnected genetic effects. A recent study published in *Molecular Psychiatry* (involving thirty-two coauthors) used the latest genome-wide association technology to scan the genomes of 3,511 people to look for differences associated with cognitive traits, including IQ—the highly heritable trait at the very center of behavior geneticists' interest.[41] The study turned up no "IQ genes," and together the 550,000 single nucleotide polymorphisms (points where the genome is known to vary by one genetic letter) explained only about 1 percent of the population variance in cognitive ability. Analyzing the data different ways confirmed the heritability estimates that low-tech twin studies provide. Though not exactly a failure, this publication represents a tremendous amount of money and effort for an extremely limited success.

The disappointment of molecular genetics and genomics in behavioral studies has been an ongoing process dogging efforts at every turn. There have been many sharp pangs, when a particularly celebrated claim has been knocked down, but this has mostly been an unfolding drama without decisive moments. This record of scientific failure has been a well-kept, open secret.

Coping with Failure

Leading psychiatric geneticist Elliot Gershon said in 1994 that the field "could not survive" another set of disconfirmations like those of the late 1980s.[42] He was wrong not only in his optimism that the hard times were over but also in his claim that they would be fatal. How did behavior geneticists manage the disappointments of molecular genetics? Some critics would call for the cessation of research.[43] And during an earlier era of controversy, when the genetics of IQ and racial differences were being debated in the 1970s, many scientists had fled the field completely. But in the molecular era, behavior geneticists produced a set of arguments and practices that pushed problems down the road while expanding their research opportunities.

One of the most basic strategies for dealing with the disappointments of molecular genetics has been to lower expectations. In the early twentieth-century prehistory of behavior genetics, it was common for scientists to assume that complex characteristics—feeblemindedness, pauperism, seafaringness—were inherited wholesale in basically Mendelian fashion.[44] Geneticists had long known this view to be false, and indeed, the application of molecular genetics was predicated on its falseness. If the patterns were Mendelian, it would have been possible to discern them through statistical analysis of family data (though not to identify their location on DNA). Still, through much of the 1990s at least, there was great hope in behavioral and psychiatric genetics that genes of relatively large effect would be found for behaviors and mental disorders. "It is obvious that these are likely to be forthcoming very soon," wrote Michael Rutter and Robert Plomin in 1997.[45]

As time has gone on, behavior geneticists have come to expect smaller and smaller effects. As several field leaders wrote in their introduction to the volume *Behavioral Genetics in the Postgenomic Era*, "Perhaps 1 gene will be found that accounts for 5% of the variance, 5 other genes might each account for 2% of the variance, and 10 other genes might each account for 1% of the variance. . . . Not long ago, a 10% effect size was thought to be small, at least from the single-gene perspective in which the effect size was essentially 100%. However, for behavioral disorders and dimensions, a 10% effect size may turn out to be a very large effect."[46] In 2003, when this was written, it would have been seen as a bold, yet mature, admission. In many cases, it might take sixteen different genes to account for only a quarter of the population variance in a behavioral trait. The field is confident enough, they seem to be saying, that this is not a problem. But by 2011, when 500,000 differences across the genome could account for only 1 percent of the variance in IQ, the humble picture from 2003 suddenly looked wildly optimistic, and behavior geneticists have had to revise expectations further downward.

Another strategy for coping with these failures has been to reconceptualize behaviors. One version of this has been to argue that many behavioral traits and mental disorders that look like a single thing at the population level are actually agglomerations of individuals expressing different causal pathways.[47] Robert Plomin, Michael Owen, and Peter McGuffin discuss this with respect to mental retardation.[48] Some diagnosed with mental retardation have the single-gene metabolic disorder PKU, others have various forms of Fragile X, and others have complex combinations of genes working in probabilistic fashion with the environment. The idea here is that "mental retardation" is actually composed of different subtypes, some genetic, some environmental, some complex combinations of both. A similar logic might apply to traits such as intelligence; from a genetic point of view there may be many different types of, say, high intelligence that produce similar scores on an IQ test. Thus, genes still matter, but they are difficult to identify in scans of aggregated populations.

In a related move, traits have also been reconceptualized as collections of "endophenotypes." Psychiatrists define mental disorders as aggregates of symptoms. As Irving Gottesman and Todd Gould describe, those diagnosed with schizophrenia might be exhibiting, among other things, poor working memory, eye-tracking dysfunction, difficulty filtering information, and exaggerated startle responses.[49] Rather than investigating the genetics of "schizophrenia" writ large, the endophenotype strategy would investigate the genetic links to such intermediate phenotypes. This strategy suggests that molecular genetics' problems are due to our aggregating too many different kinds of things under the same label. If these strategies for confronting the failure of molecular genetics work, then they will be accompanied by a dissolution or disaggregation of current behavior and disorder categories.

Technological optimism is another coping strategy. One skeptical behavior geneticist told me, "Instead of ever facing up to what hasn't worked, the field always just moves on optimistically to the next level of technology which is always available."[50] Behavior geneticists have tried to escape failure by climbing a technological ladder: At the bottom, linkage analysis looks at family pedigrees to identify genetic markers or chromosomal regions that affected family members hold in common. The technique requires few markers to scan the entire genome but has poor resolution for detecting small genetic effects. When linkage faltered, researchers climbed to the next rung and sought behavioral associations with candidate genes. This technique enables researchers to ignore family relationships and compare individuals who have a trait with controls who do not, but markers must be very close to the "candidate gene" of effect to be detected. Researchers also realized that the candidates could be wrong, and they need to look for unknown associations. Thus, one rung up, researchers have turned to GWAS that scan 500,000 to 1,000,000 differences in single nucleotides across the entire genome. As this tool has also been disappointing, geneticists have begun to look up the ladder at whole-exome sequencing (which would yield the entire part of the genome that is transcribed and translated into proteins), whole-genome sequencing (which would capture all types of genetic variation), or gene expression or epigenetic arrays (which reveal how genes are actually expressed or the inheritable changes that can regulate expression).[51]

This optimism that the next level of technology will overcome past disappointments is always accompanied by doubling down and scaling up. That is, the basic diagnosis of failure has been that researchers have looked at too few points and types of genetic variation in too few people. The response is therefore that researchers need to invest more money in higher-resolution technologies and to collect larger samples of research subjects. Recently, ninety-six psychiatric geneticists signed a collective letter asking funders not to "give up on GWAS" despite the lack of promising results.[52] They claimed that if they had funding to quadruple sample sizes, psychiatry's genetic findings would match those now available for many nonmental illnesses and traits. As a result, behavior genetics, but particularly psychiatric genetics, is becoming more like "big science," where research groups from several institutions must pool their resources (especially their research subjects) to generate publishable results with the labor and the credit distributed among dozens of coauthors (see also chap. 7 of this volume).

Postgenomic concepts have also been enlisted to explain more specific failures or anomalies in behavior genetics. For example, while monozygotic twins' relative similarity to each other has been the cornerstone of the field's science, behavior geneticists have sometimes struggled to explain twins' discordance for traits they have argued are highly genetic. Epigenetics has been offered as an explanation.[53] Perhaps the durable and inheritable environmentally acquired molecular factors that affect gene expression are making twins look less like each other than expected. Many hope that postgenomic research will help address the puzzle of hidden or missing heritability that hounds behavioral genetics (and genetic research on complex traits more broadly). Twin studies show that genetic factors influence variability in nearly every behavioral trait, yet, as I have explained, specific chromosomal loci, markers, or genes have eluded detection. Common rare variants or de novo mutations have been suggested as possible explanations: genetic factors might be making twins more similar, but they are different factors from pair to pair of twins.[54] It is hoped that bigger and more sophisticated studies could identify these factors.

Sociologists Michael Arribas-Ayllon, Andrew Bartlett, and Katie Featherstone have identified a key rhetorical strategy that psychiatric geneticists use to cope with disappointing research. They show that in review articles that appraise progress in psychiatric genetics, authors universally mobilize a discourse of "complexity" to account for the field's disappointments and failures. Complexity in these accounts refers to the "witches' brew" of multiple small genetic effects, gene interactions, gene-environment interactions, and nongenetic effects that are imagined to cause mental illnesses. They show that this discourse "incorporates criticisms, while at the same time deleting or minimizing the controversies from which they are derived."[55] Complexity talk allows the field to exonerate failure, justify intensified efforts, defend against charges of determinism, and project moderation and responsibility. What these authors do not point out is that a skeptical reading might see complexity talk as little more than cant: it can be seen less as a positive description of how an array of factors fit together than an admission of ignorance based on the failure of simple explanations. Even so, this language serves the important rhetorical function of justifying and integrating the field's practical adaptations to keep activity churning along.

An interesting feature of these failures and coping strategies is that the elements of disappointment are discussed constantly, but there has been surprisingly little rumination on the meaning of their accumulation or what would constitute general "failure."[56] One of the few behavior geneticists vocal in criticism said, "If there's any negative feeling about me in the [BGA] organization it's that people feel like things that I've written make it harder for people to get grants. Because they don't want someone saying, 'Well the truth is we're not making any progress . . . there's no big answer out there.' . . . That is not what the funding agencies want to hear."[57] At least ninety-six psychiatric researchers seem to be worried about exactly that. In earlier eras in the face of external criticism and scrutiny, behavior geneticists adopted a kind of bunker mentality that discouraged internal dissenting talk.[58] The bunker mentality here was different, one based less on defense against outside attack and more on the collective censorship that Pierre Bourdieu argues is based on intense investment in common intellectual struggle.[59]

Behavior Geneticists as Holistic Environmentalists

Even as veteran behavior geneticists eagerly pursued molecular genetics research, the competition with molecular and psychiatric geneticists made them uneasy. Behavior geneticists worried that they would not be able to beat molecular geneticists at their own game and that their research and expertise would become devalued. At the same time, the shortcomings of molecular genetics created new opportunities for the quantitative genetic techniques that had been viewed as relics. This anxiety led the veterans to hedge their bets. They reimagined their expertise as against the genetic reductionism and determinism of molecular genetics. And this filtered into their research as they reworked it to focus on measuring environmental effects and interactions rather than genetic determination.

Postgenomics is often considered more "environmentalist" than its predecessors on the basis of technological advances allowing measurement of, for example, environmentally affected gene expression and epigenetic alterations. But the postgenomic environmental turn in behavior genetics has had the opposite pattern: a return to low-tech statistical methods motivated by competition with molecular researchers and the disappointments of molecular technologies. In this volume Keller, Stevens, and Shostak and Moinester, among others, have noted the postgenomic rewriting of the "gene" and "environment" concepts. However, among the veteran behavior geneticists there was less a shift of paradigm than a shift of emphasis away from genocentrism, but within the nature/nurture framework.

An important component was to position themselves as more concerned with the phenomenon of behavior and the whole person than their molecular genetics counterparts. For example, one eminent psychological behavior geneticist marked the following distinction with psychiatric researchers: "A clinician, a psychiatrist, being a medical doctor, they tend to just totally believe in the sanctity of diagnoses. So they don't have any problem with measurement. You just follow the DSM criteria, you tick the boxes, 'Yes, they're schizophrenic.' 'No, they're not.' And they believe these things; you know, where psychologists believe behavior is much more complex and they can't—you know, a diagnosis is a very rough way of getting at this. . . . So [psychiatrists] weren't plagued by any doubts there. And doctors aren't much plagued by doubts, anyway."[60] The speaker's argument was that the behavior geneticists, coming from psychology, are much more sensitive to the complexity of behavioral traits than psychiatrists (and by extension molecular geneticists). This sensitivity, he claims, has led them to be more epistemologically reflexive. This sentiment was put more boldly by the famous behavior geneticist Thomas Bouchard when he said in an interview, "Molecular genetics looks at genes, not whole, live human beings. Twin studies add a very necessary human element to genetics and that needs to be recognized."[61] An even more radical position has been advanced by Eric Turkheimer, who has called on behavior geneticists to drop the charade that biological reduction is the aim of their work; to admit, simply, that they are social scientists; and to accept that complexity and ambiguity will always accompany their work.[62]

What should be noted is that here veteran behavior geneticists positioned themselves in precisely the opposite terms that they used in their competition with psychologists and social scientists before the molecular era. Then they saw it as a point of pride to reduce behavior to genetic causes; they dismissed careful study of the complexity and dynamics of behavior and the person. With these positions, psychologists often perceived them as crude and naïve—perceptions they came to have of molecular scientists.

Instead of antagonizing molecular geneticists, this repositioning sometimes allowed veteran behavioral geneticists to cooperate with them more effectively. Here is an example. According to several of my interviewees, the U.K. Medical Research Council's support of molecular genetics research in psychiatry and behavior was controversial in the mid-1990s precisely because of anxieties among the public and critical scientists about genetic reductionism.[63] One psychological behavior geneticist said that this controversy led to the hiring of behavior geneticists at a U.K. research center: "At a very political level, it was a hire that would counteract the public perception—or the MRC had a concern that there was a public perception that they had gone too far, too fast in the area of molecular genetics. So they needed to be able to show that there was a counterpoint to that within this center. . . . The reason we were hired here is because the MRC, after funding this center, got cold feet about its public reputation going too far in the molecular genetics direction."[64] In this climate, behavior geneticists who took behavioral measurement and environment relatively seriously could help pressured molecular geneticists soften their image. Such a perception of mutual benefit likely also helped facilitate interdisciplinary conferences and research.

This "environmental" repositioning went beyond image and had an effect on behavior geneticists' research. This same speaker put the point clearly: "I think that traditional behavior geneticists who had invested a lot of their careers and a lot of money in building twin samples, and who now have these twin samples as their major research tool, were kind of panicked by that belief that appeared with the announcement of the Human Genome [Project conclusion] that their research was worthless. And, so what I see is an increasing interest among that community of scientists who have big twin studies and adoption studies . . . using them as tools to look at environmental effects."[65] Veteran behavior geneticists' competition with molecular researchers produced anxiety about the devaluing of the resources and expertise they had accumulated. This made a new focus on the environment expedient.

Behavior geneticists had long claimed that environmental effects were an important part of their research; after all, so the argument went, if most traits are about 50 percent heritable, that means that the other 50 percent of the variance is due to environmental variance. From claims like this (mobilized mainly, it seemed, to defend against charges of genetic determinism), behavior geneticists focused closely on partitioning environmental variance into the shared and nonshared varieties. This was a period, during the early 1990s, where behavior geneticists pushed the interpretation that variations in normal parenting and other forms of socialization do not matter much. Studying the environment typically meant further and further constraining the space for traditional notions of environmental causation.

But since about 2000, as the competition intensified with molecular genetics (and chinks in its edifice became more apparent), behavior geneticists have taken environmental study more seriously. For example, psychological behavior geneticist Eric Turkheimer and colleagues conducted an important study about the dependence of IQ heritability on socioeconomic status (SES).[66] For high-SES kids they found, like many behavior geneticists before them, that IQ variance is overwhelmingly due to genetic variance; what was new was that for low-SES kids environmental variance explained nearly all of the IQ variation. Turkheimer interpreted this as showing that poor kids' environments are not stable enough for genes to matter, which argues against the view that their environments are "good enough" to let kids reach their potential.

Other examples demonstrate how behavior geneticists used their methods to clarify environmental hypotheses and effects. For one, a psychological behavior geneticist described research on the hypothesis that a father's household absence leads kids to be antisocial: "When we controlled for genetic transmission, what we find is that there is no association between father absence and children's anti-social behavior. So it appears that that's just an artifact of men with an antisocial predisposition being more willing to abandon their kids."[67] Behavior geneticists Danielle Dick and Richard Rose led a team on a series of studies on smoking and drinking behavior.[68] By comparing twins and matched nontwin controls, they found that neighborhoods and schools greatly affected these youths' smoking and drinking and that the effect of genetic differences depends greatly on qualities of the environment (in settings with easy access to alcohol or lots of peer role models, genetic differences influence behavioral differences more).

Veteran behavior geneticists, not molecular geneticists, were also the leaders in the molecular research on gene-environment interactions. The team led by Avshalom Caspi and Terrie Moffitt produced two of the most celebrated. I mentioned one earlier: the finding that the degree to which stressful experiences translate into depression depends on the form of a particular gene. The other study asked why child abuse can lead to different degrees of antisocial behavior. They found that mistreated children with the "low activity" form of the monoamine oxidase A gene exhibit much more antisocial behavior than mistreated children with the "high activity" form.[69] Behavior geneticists were attracted to this research in part because they were historically sensitive to the charges that their research promoted genetic determinism. They believed that it showed that their claim to be interactionists concerned with nature *and* nurture was more than just talk. Further, in both of these studies the effect of genes alone was not significant; genes only mattered in concert with particular experiences. Caspi and Moffitt hinted that this interdependence, seldom measured, might be a reason why molecular genetic studies had been disappointing.[70] Implicitly, this was an argument that social science–oriented behavior geneticists, because of their better grasp on environmental measurement, would beat molecular geneticists at the gene-hunting game.

It is important to realize, however, the limits to the environmentalism and anti-determinism that seems to have been induced, albeit largely indirectly, in the succession of genetic, to genomic, to postgenomic eras in research. Indeed, this is more of a shift in emphasis than a shift in paradigm. Behavior genetics has long faced a developmentalist critique that states that nature and nurture are profoundly misunderstood as distinct "factors" that can be analytically divided and separately measured.[71] Even a trait that is 100 percent heritable still involves a complex interaction between an organism's genes and environment to develop. Behavior geneticists' focus on environmental factors and interactionism has involved looking at different parts of the nature versus nurture equation, not a rethinking of the presumptions of that equation or the notion of the analytic separability of genes and environment.

Postgenomics has been incorporated in a similarly intellectually reproductive way. Critic Evan Charney has argued that postgenomic research has discredited the behavior genetic paradigm from multiple directions: for example, epigenetic regulation, DNA variability, and somatic mosaicism, particularly prevalent in the brain, undermine assumptions undergirding the field's dominant methodologies.[72] However, it seems that more often postgenomics has been incorporated into the existing research paradigms. For one, behavior geneticists' carefully collected twin and family pedigrees and data have gained new currency with researchers eager to apply postgenomic technologies. But more to the point, when epigenetics is invoked as a reason for twins' discordance, the presumption is that genes would have determined their similarity if not for epigenetic interference.[73] Similarly, the idea of the ubiquity and efficacy of rare genetic variants influencing behavior does not push beyond genetic determinism, but rather suggests that there are multiple genetic determinisms that are difficult to detect in genetically heterogeneous populations (see also chap. 11 of this volume).

Conclusion

So what does this case study of behavior genetics tell us about postgenomics? Though the narrative of postgenomics is still very much in formation—as the present volume attests—there are certain recurring features. A certain intellectual ethos, driven by technological breakthroughs in both molecular biology and informatics, animates postgenomic science: researchers emphasize causal complexity and emergent properties, nondeterminism, and new kinds of gene-environment interactions. Postgenomics represents a decisive break with the triumphalist reductionism of the genetic and genomic eras—postgenomic scientists are embarrassed by the "Holy Grail" and "Book of Life" talk used to sell the HGP.

Behavior and psychiatric geneticists, I have shown, have adapted something of this ethos—emphasizing interactionism and complexity and shying away, in some contexts at least, from talk of finding "genes for" behavioral traits. But they have arrived at this point not by ascending an ever-improving technological ladder but through a combination of technological failures and games of professional distinction. Though they are undeniably enthusiastic about what postgenomic technologies hold for their field, their embrace of its ethos is based almost entirely on the shortcomings of genomic technologies and the stability of pregenomic quantitative genetics.

As with other "posts," the idea that postgenomics represents a fundamental break with older styles of research and reasoning relies on selective views of both history and the contemporary scene.[74] Postgenomics warrants arguments against determinism and nature/nurture separability.[75] But these same critical arguments, made in different empirical terms, have *always* accompanied behavior genetics.[76] Postgenomic research has not eclipsed other research paradigms in behavior genetics. Conventional molecular genetic research looking at candidate gene variants and quantitative trait loci abounds, as does pre-molecular quantitative genetics. Thus, rather than a decisive break with the old, postgenomics in behavior genetics represents a proliferation of coexisting and interacting scientific possibilities.

Nor has the advent of postgenomics signaled an era of caution and reserve at the intersection of molecular biology and the politics of racial behavioral differences. Indeed, it is new experts who have spoken with the greatest recklessness and abandon. The recent provocative public claims about purported genetic differences among races in behavior or intelligence have come not from veteran behavior geneticists but from newcomers such as Bruce Lahn, James Watson, and Henry Harpending.[77] Rather than setting the conditions for greater integration and policing among scientists, postgenomic technologies seem to be exacerbating the fragmented character of the scientific community and have not stimulated open conversations about scientific standards.[78] Rather than an era of consensus, postgenomics has been characterized by the proliferation of controversies and the reinforcement of old professional and conceptual divides.

This case suggests that closer attention be paid to the role of failure in dynamics of scientific change. The failures of technologies to deliver expected results have helped drive scientists to alternate technologies, yet this process has been decidedly non-Popperian. First, failures have not led cognitive frameworks to be abandoned, but rather led different niches in them to be emphasized. Gene/environment partitioning does not go away, for example, but the focus shifts to environmental effects. Second, scientific disappointments have tended to spur calls for expanding and intensifying research—doubling down rather than folding. They have further encouraged returns to lower technological levels and their relative stability. Thus, rather than withering away, scientific failures have tended to spur research and draw resources. These disappointments may have been a blow to the field's cognitive authority, but they have produced a boom in practical activity and investment. For all the accomplishments of genetics, has there ever been a field so lavishly rewarded with attention and resources despite consistent failures to deliver on promises?[79]

Mike Fortun (chap. 3 of this volume) proposes an affective history of genomics. My narrative suggests that to the pleasure and excitement of surprising results we must add the disappointment and anxiety of unmet expectations. Are such emotions mostly local, linked to the fate of the individual experiment or research team? Perhaps the welcome surprise becomes a collective resource that can help those anxious of failure motivate and justify future efforts.

Finally, it is worth reflecting briefly on the unintended scientific consequences of postgenomics. Anthropologist Stefan Helmreich has written about evolutionary biologists' efforts to incorporate genomics in order to make their taxonomical schemes more unified and rigorous and to resolve previously obscure evolutionary relationships among life-forms.[80] Instead of securing taxonomy, genomics undermined the very idea of taxonomy and threw the concept of evolutionary relatedness into doubt; rather than a stately tree, evolutionary relationships (especially of microorganisms) have become a rhizomatic tangle. The failures of the molecular in behavior genetics may have similar unintended consequences and transform the way we think about behavior, genes, and environment. But it may just undermine the imperial ambitions of genetics. As one skeptical behavior geneticist put it, "The impact of genetics on the overall undertaking of behavioral science has been and will be less than everybody thought it was going to be. And that in fact is what happens when the traditional problems in psychology meet a new technology, it's not that the technology transforms the old problems into new science, it's that the complexity of the problems transforms the technology into social science of the old kind."[81] Perhaps genomics will be more affected by its encounter with behavioral science than the reverse. Surely, this is but one more prediction, albeit one that is not tacking upwind.

Notes

Parts of Chapter 8, "From Behavior Genetics to Postgenomics" by Aaron Panofsky first appeared as part of his book, *Unbound Science, Fragmented Knowledge: Controversies and the Development of Behavior Genetics*. These parts are reproduced by permission of the publishers, University of Chicago Press © 2014. All rights reserved.

1. Rogaev, "Genomics of Behavioral Diseases."
2. Lewontin, "Analysis of Variance and the Analysis of Causes."
3. Plomin et al., *Behavioral Genetics in the Postgenomic Era*; see also Billings, Beckwith, and Alper, "Genetic Analysis of Human Behavior"; Joseph, "Crumbling Pillars of Behavioral Genetics."
4. Alper and Beckwith, "Genetic Fatalism and Social Policy"; Duster, *Backdoor to Eugenics*, and "Selective Arrests"; Nelkin and Lindee, *DNA Mystique*, 2nd ed.; S. Rose, "Neurogenetic Determinism and the New Euphenics."
5. Panofsky, *Misbehaving Science*.
6. Beckwith, "Foreword," 1–2.
7. Kevles, *In the Name of Eugenics*.
8. Gilbert, "Vision of the Grail," 86.
9. Quoted in Jaroff, "Gene Hunt."
10. Watson, "Molecular Genetics Perspective," xxii.
11. Koshland, "Nature, Nurture and Behavior," "Sequences and Consequences of the Human Genome," and "Rational Approach to the Irrational."
12. Hamer and Copeland, *Science of Desire*, 25.

13. I obtained the data for fig. 8.1 using a keyword search for "behavioral genetics" (the NIH's database label for the field) on new (nonrenewal) research grants (eliminating other forms of funding support) in the NIH's now-defunct "CRISP Database" (http://crisp.cit.nih.gov/). Funding data are difficult to obtain for scientific subfields (especially when keywords associated with the subfield are not highly specific).

14. NIH has historically not provided funding data with the CRISP database. (This is changing with the new RePORT system.) These funding data were obtained from an online resource from an advocacy group called the Sunshine Project. They combined NIH's budgetary disbursement database with its grant database between 1996 and 2005 in an online resource called "CRISPer"; the project, now defunct, was available at www.cbwtransparency.org/crisper.

15. The organization was originally called the European Behavioural and Neural Genetics Society, but to broaden its reach, "International" was added to the name (and "European" removed) at the first conference in 1997. See http://www.ibangs .org/index.php?option=com_content&view=article&id=33.

16. Membership rosters were published in 1985 and 1996; aggregate numbers are published in the BGA's minutes in *Behavior Genetics*.

17. This is a ballpark estimate. Circa 2011–12, BGA membership was down to 444, ISPG was 550, and IBANGS was 228. There is some overlap between the societies, so membership cannot simply be summed.

18. Panofsky, "Field Analysis and Interdisciplinary Science."

19. Kendler, "Discussion," 100.

20. Panofsky, "Field Analysis and Interdisciplinary Science," and *Misbehaving Science*.

21. Kaplan, *Limits and Lies of Human Genetic Research*; D. Moore, *Dependent Gene*; S. Rose, "Rise of Neurogenetic Determinism."

22. Egeland et al., "Bipolar Affective Disorders."

23. Kelsoe et al., "Re-evaluation of the Linkage Relationship."

24. Robertson, "False Start on Manic Depression."

25. Risch and Botstein, "Manic Depressive History," 351.

26. Kato, "Molecular Genetics of Bipolar Disorder and Depression."

27. Hamer et al., "Linkage between DNA Markers."

28. Miller, "Introducing the 'Gay Gene.' "

29. Hamer and Copeland, *Science of Desire*.

30. G. Rice et al., "Male Homosexuality"; see also Mustanski et al., "Genome-wide Scan of Male Sexual Orientation."

31. Blum et al., "Allelic Association of Human Dopamine d2 Receptor Gene in Alcoholism."

32. Holden, "Cautionary Genetic Tale."

33. Conrad and Weinberg, "Has the Gene for Alcoholism Been Discovered Three Times since 1980?"

34. Caspi et al., "Influence of Life Stress on Depression."

35. News Editorial Staff, "Runners-up."

36. Risch et al., "Interaction between the Serotonin Transporter Gene (5-Httlpr), Stressful Life Events, and Risk of Depression."

37. Zammit, Wiles, and Lewis, "Study of Gene-Environment Interactions in Psychiatry"; Zammit, Owen, and Lewis, "Misconceptions about Gene-Environment Interactions in Psychiatry."

38. Hirschhorn et al., "Comprehensive Review of Genetic Association Studies."

39. Bosker et al., "Poor Replication of Candidate Genes"; Buxbaum, Baron-Cohen, and Devlin, "Genetics in Psychiatry"; Charney and English, "Candidate Genes and Political Behavior"; Newton-Cheh and Hirschhorn, "Genetic Association Studies of Complex Traits"; Siontis, Patsopoulos, and Ioannidis, "Replication of Past Candidate Loci"; Sullivan, "Spurious Genetic Associations."

40. Gorroochurn et al., "Non-replication of Association Studies"; van den Oord and Sullivan, "False Discoveries and Models for Gene Discovery."

41. Davies et al., "Genome-Wide Association Studies"; see also Bouchard, "Genetic Influence on Human Intelligence (Spearman's G)"; Deary, Johnson, and Houlihan, "Genetic Foundations of Human Intelligence."

42. Marshall, "Highs and Lows on the Research Roller Coaster," 1694.

43. Joseph, *Missing Gene*.

44. Kevles, *In the Name of Eugenics*.

45. Rutter and Plomin, "Opportunities for Psychiatry from Genetic Findings," 214.

46. Plomin et al., "Behavioral Genetics," 9.

47. Scott F. Stoltenberg and Margit Burmeister call this "equifinality"—where different causal paths can produce the same outcome; Stoltenberg and Burmeister, "Recent Progress in Psychiatric Genetics," 927.

48. Plomin, Owen, and McGuffin, "Genetic Basis of Complex Human Behaviors."

49. Gottesman and Gould, "Endophenotype Concept in Psychiatry."

50. Personal interview (#35, June 2004).

51. Rogaev, "Genomics of Behavioral Diseases."

52. Sullivan, "Don't Give Up on GWAS."

53. Wong et al., "Longitudinal Study of Epigenetic Variation in Twins."

54. Charney, "Behavior Genetics and Postgenomics"; Taylor, "Three Puzzles and Eight Gaps."

55. Arribas-Ayllon, Bartlett, and Featherstone, "Complexity and Accountability," 516–17.

56. Behavior geneticist Eric Turkheimer stands out for his meditations on failure; see Turkheimer, "Mobiles," and "Commentary." So does psychiatric geneticist Kenneth Kendler (Kendler, "Psychiatric Genetics"), though he offers a slightly more optimistic vision.

57. Personal interview (#35).

58. Panofsky, *Misbehaving Science*.

59. Bourdieu, *Science of Science and Reflexivity*, 62–71.

60. Personal interview (#28, February 2004).

61. Miele, "En-twinned Lives," 37.

62. Turkheimer, "Mobiles," and "Spinach and Ice Cream."

63. Personal interviews (#24, February 2004; #25, February 2004; and #30, February 2004).

64. Personal interview (#25, February 2004).

65. Ibid.

66. Turkheimer et al., "Socioeconomic Status."

67. Personal interview (#25, February 2004).

68. Dick et al., "Exploring Gene-Environment Interactions"; Dick and Rose, "Behavior Genetics."

69. Caspi et al., "Role of Genotype."

70. Caspi et al., "Influence of Life Stress on Depression," 389.

71. Gottlieb, "Some Conceptual Deficiencies"; Keller, *Mirage of a Space*; Lewontin, "Analysis of Variance and the Analysis of Causes"; Moore, *Dependent Gene*.

72. Charney, "Behavior Genetics and Postgenomics."

73. Indeed, Charney argues here that epigenetics can have a variety of effects, including increasing concordance, but behavior geneticists have been mainly interested in just the cause of discordance.

74. See Calhoun, *Critical Social Theory*, chap. 4.

75. Charney, "Behavior Genetics and Postgenomics."

76. See, e.g., Hirsch, "Behavior-Genetic, or 'Experimental,' Analysis"; Lewontin, "Analysis of Variance and the Analysis of Causes"; Lewontin, Rose, and Kamin, *Not in Our Genes*; Moore, *Dependent Gene*.

77. Mekel-Bobrov et al., "Ongoing Adaptive Evolution of Aspm"; Hunt-Grubbe, "Elementary DNA of Dr Watson"; Cochran, Hardy, and Harpending, "Natural History of Ashkenazi Intelligence."

78. Panofsky, "Rethinking Scientific Authority"; Richardson, "Race and IQ in the Postgenomic Age."

79. We label a sports team a failure when it posts a losing record, but we usually forget to ask whether it earned a profit for its owners. The analogy suggests that we should think about conceptual as well as practical/material dimensions of success and failure in science.

80. Helmreich, "Trees and Seas of Information."

81. Personal interview (#35, June 2004).

9

Defining Health Justice in the Postgenomic Era

Catherine Bliss

The concept of health disparities, a central feature of the postgenomic understanding of health justice, was once absent from genomic discourse. The concept emerged amid public health debates over the underrepresentation of women and minorities in U.S. research in the mid-1980s. While activists, policy makers, and health experts grappled with how to bring about greater inclusion of underrepresented groups, geneticists were completely independently honing the methodologies and technologies that would become the field now known as genomics. Even the public health subfield of genetic epidemiology remained absent from debates over racial inclusion. When the National Institutes of Health (NIH), the U.S. Department of Energy (DOE), and the world's leading human genome research institutes launched the Human Genome Project (HGP) in 1986, the project took no notice of minority health or population-based inequalities. Yet, today postgenomic sciences are leading a movement in the biosciences to address health inequality. Their sponsoring institutions for health justice and funding mechanisms hold the largest budgets for health disparities research. Health justice is increasingly approached through next-generation genome sequencing and gene-environment frameworks.

This chapter explores this rapid reversal by demonstrating how the parallel rise of DNA solutions and equity campaigns in public health has brought postgenomics to bear on matters of health justice. Analysis of public health policies, genome project records and news coverage, and interviews with leading postgenomic geneticists will show how studying health disparities became a political necessity for life scientists just as genomics was moving into the more environment-conscious postgenomic era. Genome mappers first moved from population-blind sampling protocols to continental ancestry-based, minority-conscious sampling protocols. They then pinned the reputation of the emerging postgenomic field on its racialized technological and bioethical solutions. As government bodies such as the U.S. Public Health Department have ramped up investment in postgenomic research models across the life sciences, they have also increasingly brought a postgenomic approach to already-defined epidemiological research areas focused on health disparities.

This chapter begins by documenting the biopolitical shift in policies and protocols that made DNA research a guiding framework, and health justice a requisite concern, for the life sciences, paying specific attention to their influence on research strategies and priorities across the genomic sciences and within large-scale sequencing efforts. Next, it presents how leading geneticists came to conceptualize and popularize health justice as a postgenomic concern, while putting forth postgenomics as a frontline weapon against inequality. The chapter then examines the implementation of postgenomic frameworks in the context of public health and disparities research. Finally, it examines evidence of an emerging knowledge paradigm, what I call a "postgenomic style of thought," among today's elites. I interpret these events through a rubric of biopower and the sociology of knowledge, highlighting the interdependence of science and politics in a postgenomic age.

Research as Governance

Advances in the genetic sciences signal a larger social trend toward biopolitics, "the endeavor, begun in the eighteenth century, to rationalize the problems presented to governmental practice by the phenomena . . . health, sanitation, birthrate, longevity, race."[1] As states "brought life and its mechanisms into the realm of explicit calculations and made knowledge/power an agent of transformation of human life," the biosciences rendered life's most basic processes intelligible.[2] The relationship between scientific knowledge and state power has been dialectical. Governing bodies sponsor certain forms of research that shape how knowledge is collected, what can be known, and how populations are governed. Since the mapping of the human genome, governance domains have proliferated as a series of "omics" to stipulate everything from our cellular material to the ambient environment and its social impact, thus making DNA solutions a research priority in governance systems around the world.[3] Changes within the life sciences accelerate and shape the nature of biopolitical governance, just as they create dominant "grids of intelligibility" in science and society.[4]

Since 2007, health disparities research and postgenomic innovation have reigned as two of the U.S. federal government's five public health priorities.[5] Inclusion of research subjects by census racial/ethnic classifications and genetic analysis of participant groups have been the chief goals. As sociologist Steven Epstein argues, these foci constitute a biopolitical paradigm—"[a framework] of ideas, standards, formal procedures, and unarticulated understandings that specify how concerns about health, medicine, and the body are made the simultaneous focus of biomedicine and state policy."[6] Research standards wedding health justice to postgenomic science simultaneously articulate state goals and strategies and stipulate "the very nature of the problems they are meant to be addressing."[7]

Though formulated under American auspices, these standards shape a global scientific order. First, the U.S. federal government funded the bulk of genomic research during genomics' turn to postgenomics.[8] Second, the NIH was the primary sponsor and director of all the international genome projects into the early 2000s. As a result, the United Kingdom has sponsored census-based inclusion.[9] Indeed, many American policy shifts can be witnessed in U.K. Department of Health and Medical Research Council undertakings, as well as European Union governance. In the private sector, Western pharmaceutical outsourcing has created an uneven politics of recruitment where citizens of the global South present themselves as amenable to Western frameworks.[10] Research into the establishment of biotechnologies that serve racial minorities has similarly shown that U.S. researchers superimpose American standards onto foreign sites.[11]

Regarding individual scientific investigations, it is instructive to ask, "Why that particular question?"[12] How does a project reflect assumptions and standards serving the broader policy context? Yet, taking emergent research fields as collective entities, biopolitical paradigms can also be analyzed in terms of "thought disciplines," modes of "identifying difficulties, questioning arguments, identifying explanatory failures."[13] In newly autonomizing fields, where a few scientists dominate research agendas, expert perspectives can elucidate the penetrance and valence of a particular mode of governance.[14] In the formation of postgenomics, these leading scientists assimilated the dominant political framework, one focused on a politics of ethnic inclusion and diversity, in part because it aided the field in developing a positive ethical valence with the public. As a result, postgenomic scientists invested in the notion that genomics would lead the world out of its predicament with racial inequality.

Health Justice in the Genomic Age

"Health justice," the watchword of feminist and antiracist political campaigns that created inclusionary measures across the public domain,[15] and "health disparities," the conception of health justice as a research area in its own right,[16] emerged alongside the incipient field of genomics in the early 1980s.[17] By 1986, the idea that certain populations were "vulnerable" in biomedicine and health care had been institutionalized in the first NIH-wide policy corrective to disparities affecting women and racial minorities. By 1986, the first genome project had also launched. Initially, there was no overlap between these fields.[18] Yet, as a tacit coalition of policy makers, activists, and biomedical experts devised ways to level the biomedical playing field,[19] genomic research and development would come to accept health disparities as a central concern.

The initial absence of cross-pollination is most clear in the trajectories of the international human genome projects of the late twentieth century. The HGP, launched by the NIH and DOE in 1986, made little mention of health disparities during its first decade of existence. Even as its bioethical innovations, such as its Ethical, Legal, and Social Issues (ELSI) branch, were touted as the new paradigm in public health, health disparities did not factor into its program until 1999.[20] When the HGP reformulated its goals midway through the project, the language of health disparities was absent from its list of public health concerns.[21] Conceptually, the emergent field saw all DNA, and all research subjects, as being the same.[22] Project leaders pooled DNA from random sources to sequence contiguous exomes.

Debates in public health would eventually shift the hand of NIH policy makers as the first postgenomic technologies were being conceived. In 1990, the Office of Research on Women's Health and the Office of Minority Programs were established at the NIH. That year, the Surgeon General's Office issued its decennial Healthy People plan with a call to "increase the span of healthy life," "achieve access to preventive services," and "reduce health disparities . . . among all Americans."[23] The Surgeon General's Office simultaneously embarked on a plan to implement federal race standards in national health statistics. In 1993, the NIH issued the Revitalization Act of 1993, a statute on the surveillance and inclusion of women and racial minorities in clinical research.[24] The Centers for Disease Control and Prevention (CDC) also released the "Use of Race and Ethnicity in Public

Health Surveillance," a statement affirming health disparities as a leading issue for public health.[25]

Amid these debates, DNA research took hold on a global scale.[26] An international team of researchers launched the Human Genome Diversity Project, a second project that would compare populations across the world. From the outset, the Human Genome Diversity Project promised a more active inclusionary policy, including a first look at the genomes of non-"Caucasoid" people.[27] Despite federal sponsorship from numerous agencies, including the DOE, National Science Foundation, and National Institute of General Medical Sciences, the project did not align itself with federal efforts to redress health disparities. Rather than operate with a clear program for the inclusion of women and U.S. racial minorities, the project used ethnic and tribal affiliations to guide population sampling.[28]

As the project got under way, indigenous polities from around the world protested the use of their DNA. In 1995, under mandate of the U.S. Department of Health and Human Services (HHS), the National Research Council investigated the Human Genome Diversity Project's plans for recruiting vulnerable populations. Project leaders responded by launching a campaign to include U.S. racial minorities.[29] In the end, the National Research Council was unconvinced that the Human Genome Diversity Project would handle minority inclusion properly, and the project went unfunded. Yet, the controversy made health disparities a priority concern for future projects.[30]

Signs of a convergence in health disparities and genomic research arose in 1996, when the HHS issued a series of policies and programs to establish health disparities research across biomedicine, specifically targeting research areas in the genetic sciences. The DOE and National Human Genome Research Center began by mandating an active inclusionary policy for women in all genetic research.[31] The policy maintained that, although male and female nonsex chromosomes were not different enough to merit sex-specific analysis, the persistence of gender health disparities demanded a concerted effort to bring women into the fold of research. The National Human Genome Research Center, which was in the process of becoming an institute of the NIH, also hosted a series of conferences on minority inclusion. These conferences sought partnerships with prominent minority organizations that could devise equitable research subject recruitment strategies.

In 1997, the newly minted National Human Genome Research Institute (NHGRI) became a scientific and bioethical beacon for all NIH institutes when it launched the first genome project to use federal racial classifications to oversample minority populations.[32] The Polymorphism Discovery Project recruited and subsequently packaged and distributed the DNA samples to researchers according to public health inclusionary mandates. The NHGRI simultaneously issued its first funding mechanism on health disparities implications of genomic research. "Concepts of Race, Ethnicity, and Culture: Examination of the Ways in Which the Discovery of DNA Polymorphisms May Interact with Current Concepts of Race, Ethnicity and Culture" not only fostered original research into the social implications of population genomics but also led policy makers to form an in-house Race and Genetics Working Group. At millennium's turn, the NHGRI stated a new goal for the HGP: investigation into "how socioeconomic factors and concepts of race and ethnicity influence the use, understanding, and interpretation of genetic information, the utilization of genetic services, and the development of policy."[33] These efforts propagated a DNA-equity paradigm in the life sciences. The health disparities turn within the world's leading genomic research institute secured its manifestation in research and development across the field.

Geneticists Define Health Justice

In the early years of the new millennium, the scientists who have since seen genomics through the era of large-scale sequencing into gene-environment frameworks and next-generation sequencing went from downplaying the necessity of genomics in solving racial health disparities to heralding its central role.[34] Initially, elites of the genome projects maintained that since gender or race have "no scientific basis," science in the wake of the genome era could move past it.[35] Yet, by the culmination of the last global genome project, elite scientists advancing the superiority of DNA research in public health would refashion themselves as leaders of disparities research. By the end of the first decade of the twenty-first century, amid a growing arena of postgenomic technologies, a geneticist would be chief of health justice policy for the entire field of U.S. public health.

Pharmacogenomics was the first postgenomic research branch to take a strong hand in health justice. Following the U.S. Food and Drug Administration's (FDA) mandate to include women and minorities in all clinical trials, drug developers such as Genaissance Pharmaceuticals's Gualberto Ruaño promoted the inclusion of minority subjects in research as a way to generate personalized medicine for underserved groups. Ruaño argued that the new genetics would show that "efficacy could be proven in small cohorts" instead of populations in the thousands.[36] Pharmacogenomic studies and reports quickly took up the language of "race-gender" variation in drug metabolism and clinical trials research.[37] Though the pharmacogenomics health disparities emphasis was not felt immediately in other drug-making realms of postgenomics such as proteomics and metabolomics, it led to the establishment of race-based therapies such as BiDil.[38] The FDA's approval of drugs and diagnostics that target specific races and genders, in the interest of health justice, provided the context within which scientists would fashion postgenomic priorities.

Geneticists who spearheaded the structural DNA analysis that has given rise to epigenetics and copy number variant sequencing also came to laud health justice as a central problematic for genetic sciences. Once the draft map of the human genome had been published, the NHGRI leadership decided to base the final global project of the genome era, the International HapMap Project, on federal race standards.[39] As Eric Lander, founding director of the Broad Institute and Millennium Pharmaceuticals, and a leader of the Human Genome and HapMap Projects, maintained, "We must make sure the information is not used to stigmatize populations. But we have an affirmative responsibility to ensure that what is learned will be useful for all populations. If we shy away and don't record the data for certain populations, we can't be sure to serve those populations medically."[40] Lander's strategy was paradigmatically aligned with projects such as a $33 million NIH heart disease study and a $22 million cancer study focused on African Americans, forming a network of public health research into genomic causes for health disparities.[41]

As geneticists moved from single nucleotide polymorphism and haplotype mapping to high-throughput functional genomics, the federal government fortified its health justice policies and its reliance on DNA solutions. In 2000, the NIH mandated minority community consultation in all genetic research.[42] It also instituted institute-specific Strategic Research Plans to Reduce and Ultimately Eliminate Health Disparities that would fund genomic and postgenomic research in the study of racial disparities.[43] These were massive amounts of funding that would geneticize all public health institutes. As the NIH committed $1.3 billion to funding trans-institute research into health justice, the HHS released Healthy People 2010 and the Initiative to Eliminate Racial and Ethnic Disparities.[44]

Amid these events, geneticists put a health justice imprimatur on translational genomics. Members of the NHGRI-sponsored Pharmacogenomics Research Network, as well as their colleagues at Stanford University, voiced support for the use of federal inclusionary classifications, stating, "There is great validity in racial/ethnic self-categorizations, both from the research and public policy points of view."[45] Moving beyond a genetic epidemiology that they asserted was based on monopopulational, monogenetic inquiry, they argued that "we need to value our diversity rather than fear it, and ignoring race will hurt epidemiological assessment of disease in the very minorities the defenders of political correctness wish to protect."[46] These leaders in translational medicine argued for their colleagues to use disparities as "starting points for further research."[47]

Meanwhile, the NHGRI leadership formulated a project that would sequence one thousand whole genomes under the health justice sampling protocols established by the International HapMap Project. During project planning, a number of scientists, such as Steering Committee cochair David Altshuler, made statements supportive of using common racial classifications as a "last resort" in a proactive health justice science.[48] Amid the launch of a second postgenomic genome project, ENCODE, then–NHGRI director Francis Collins argued that geneticists had a responsibility to do whatever it took to foreground the study of racial health disparities: "We need to try to understand what there is about genetic variation that is associated with disease risk, and how that correlates, in some very imperfect way, with self-identified race, and how we can use that correlation to reduce the risk of people getting sick."[49] In fact, Collins suggested that genomics should lead the sciences in health disparities research: "I think our best protection against [racist science]—because this work is going to be done by somebody—is to have it done by the best and brightest and hopefully most well attuned to the risk of abuse. That's why I think this has to be a mainstream activity of genomics, and not something we avoid and then watch burst out somewhere from some sort of goofy fringe."[50] The leaders of the 1000 Genomes and ENCODE projects called for the new postgenomics to be at once a leader in efforts to define health justice. The formation of the biopolitical paradigm was complete.

Postgenomic Priorities

In the past ten years, NIH research budgets have been trained toward investigation into microRNAs, gene-environment interactions, epigenetics, and translational medicine. Across fiscal reports and plans, the NIH has listed the determination of the role genes play in diseases that exhibit disparities as the top priority of trans-institute funding.[51] In the turn from mapping to function, the "postgenomicized" NIH has sponsored new funding mechanisms and research centers across the United States. It has also fortified intramural research around the health justice paradigm.

In 2003, the NIH introduced an institutional award for the Centers for Population Health and Health Disparities program. By 2008, five centers had been funded and ten more were scheduled for funding.[52] The program's charge advances postgenomics as fundamental to disparities research: "to explain how the social and built environments impact biological processes, such as epigenetic modifications, gene expression, endocrine function, inflammation, tumor growth, and cancer-related health outcomes. These types of information are crucial in developing appropriate prevention, early detection, and treatment intervention programs to mitigate cancer disparities."[53] To access these multimillion-dollar awards that exceed any other level of social science funding, researchers must marshal a host of gene-environment technologies.

Also in 2003, the NHGRI, DOE, and National Institute for Child Health and Human Development issued an institutional award for Centers of Excellence in Ethical Research. Though these centers are required to further a range of ELSI policies and programs, they were initially envisioned as hubs of gene-environment health justice research. Funders intended the centers to partner with minority institutions, including race-based associations and historically black universities. The NHGRI 2004–8 Health Disparities Plan announced the then-reigning inclusionary framework: "Special emphasis will be placed on access to information, informed consent, community attitudes toward genetic research; emphasis will also be placed on the development of methods to optimize informed decision-making regarding participation in genetic research and use of the knowledge gained through this research. It is hoped that the research supported by this initiative will increase information available to investigators that will help them to design future genetic research in a way that will more successfully involve minority communities."[54] The program initially provided over $20 million in grants to study issues such as breast cancer and asthma in people of African descent, and it has since funded centers focused on gene-environment research into diabetes, prostate cancer, and sickle cell anemia in minority communities.

The U.S. government has also launched two intramural federal research centers entirely focused on health disparities gene-environment research. In 2007, the NIH established the Intramural Center for Health Disparities Genomics. Though the center has been since renamed the Center for Research on Genomics and Global Health, it is known for its steady emphasis on the health of people of African descent. The CDC also established an Office of Public Health Genomics "to convey the importance of engaging communities, investing in [community-based public research] and ensuring that social justice be central to public health genomics."[55] It has maintained race and gender stratification as a focal point of gene-environment inquiry.

Health justice is equally marshaled in the foundation of innovation agendas. Today, trans-institute projects must include a health disparities component to its gene-environment and epigenetic programs. The NIH Task Force on Obesity, for example, cites health disparities epigenetics and intrauterine interactions as its new aims (see chap. 11 of this volume). Similarly, in its effort to promote a developmental biology approach to gene-environment research in noncommunicable chronic disease (NCD) research, the NIH Common Fund, the central trans-institute administration that was initially launched to deliver postgenomics funding to all institutes, states, "Disadvantaged populations may experience greater exposure to [environmental] hazards and exhibit higher rates of disease incidence, morbidity and mortality. Understanding and modulating this risk in humans during critical windows of development offers the promise of primary prevention for many of these NCDs and may result in reducing health disparities."[56]

Indeed, the Common Fund has stipulated a health justice approach to the formulation of the Synthetic Cohort for the Analysis of Longitudinal Effects of Gene-Environment Interactions, which began as the first postgenomic-era proposal for a national prospective cohort study by Francis Collins in 2004. The NIH targets three diseases for health disparities gene-environment research: diabetes, hypertension, and prostate cancer. Work with the FUSION, FBPP, and CGEMS studies has led to the development of whole-genome sequencing research components.[57] The Gene-Environment Initiative, initiated by the HHS in 2006, holds a mirrored approach.[58] Its NHGRI-led branch, GENEVA, maintains "pathways to disparities in health outcomes" as one of its three aims.[59]

Finally, the global health domain, in which health disparities are framed as a genomic divide, has put forth postgenomics as critical to the armamentarium in the fight for equality. Following the mandate of President Obama's Global Health Initiative, the NIH Global Health Research Meeting has claimed gene-environment research as its frontline research approach. The meeting's team of sixty world-leading scientists lists "RNAi, small molecule screening, genomics of pathogens, and vaccine development" as the advances that make "an attack on infectious diseases possible."[60] With microbiomics and epigenetics, they identify health disparities research as one of their "priority areas." The first international sequencing project of the postgenomic era, the Human Heredity and Health in Africa Project, has taken gene-environment interactions and health justice as its axioms. Launched in 2010 by the NIH and Wellcome Trust, the project investigates attempts to use functional genomics to reduce the communicable and noncommunicable disease burden in populations of African descent. A 2012 conference, sponsored by planners at the NIH Center for Research on Genomics and Global Health, designated epigenomics as the latest focus of African diaspora population science. These policies signal the institutionalization of the health justice paradigm beyond U.S. borders.

A Postgenomic Style of Thought

In the world of postgenomic projects and studies, health justice leadership is the watchword of the day. In interviews conducted in 2007 and 2008, scientists revealed the ways disparities research agendas discipline their thought. Speaking on the NIH gene-environment prospective cohort study, Francis Collins argued, "From a perspective of social justice which I think most scientists would have liked to adhere to, we have to take this as one of our highest priorities to get this part right. I also have, because of being the director of this institute now for 14 years, a lot of very wise advisors around me and this regularly comes up in almost every conversation about advances that are occurring, about the need to focus on this issue, the need not to just hope it turns out right."[61] Collins maintains the health justice motto and has transferred these research priorities to the NIH trans-institute administration, ensuring their uptake in individual institutes.

Charles Rotimi, the leader of the NIH Center for Global Genomic Health and the Human Heredity and Health in Africa Project, also regards health disparities research as the geneticist's central charge in the postgenomic era. After discussing his early interest in health disparities in populations of African descent, he explained how such research led him toward a gene-environment approach: "Initially, the way we were thinking about it really was to hold genes constant and look at the environment. So by looking at people who share various ancestries, do they appear in degrees of high admixture populations? So it was sort of a very crude way of holding genetics constant, and looking at the environmental impact. That was the initial design, but of course over time it became a gene-environment study."[62] With Collins and other luminaries such as 2011 American Society of Human Genetics president Lynn Jorde, Rotimi publicizes in the leading science media venues of the postgenomic era the message that the "environment" in gene-environment is a health justice issue for postgenomic geneticists. From his post as president of the African Genetics Society, Rotimi has also been a staunch advocate of population-based models in non-U.S. research.

Even in the realm of private sector next-generation sequencing, where public standards play less of a role, innovators in postgenomics work within a health justice paradigm that is focused on race-identified genetic data and sampling. Speaking on the importance of collecting data on self-identified race and ethnicity, David Altshuler argued, "If we don't measure it then we'll never know that exists and we won't know what we needed to be looking for and also we'll miss the fact there is health disparity."[63]

Though Altshuler initially criticized mainstream media sequencing events that targeted vulnerable populations, he and his associates at the Broad Institute eventually assumed a leadership role in media events targeted toward racial minorities. In 2010, the Broad Institute teamed up with the Personal Genome Project and minority justice advocate Henry Louis Gates Jr. to publicize the importance of whole-genome sequencing for African American health. Gates and his father, the first African Americans to have their whole genomes sequenced, made numerous public appearances, introducing the concept of the Genomic Divide between racial haves and have-nots. They also convened the first Genomes, Environments, and Traits Conference to find ways to characterize the relationship between the genome, envirome, proteome, microbiome, and transcriptome for minority populations. The Broad Institute now provides next-generation sequencing for Gates's genealogy television show *Faces of America*, while Gates steers a pharmacogenomics company dedicated to creating race-based drugs that treat sickle cell anemia. In private sector advances of health justice, we see the crystallization of a thought discipline. Health justice is increasingly understood as not only a public health but a private health matter. With the range of basic research investigations, personal genomics and personalized medicine advance under a health disparities research framework.

The postgenomic style of thought is first and foremost a way of constructing bioscience as a salve for matters social and biological, with scientists at the lead. Geneticists fashion themselves as public reformers, stewards of justice. They maintain that it is their responsibility to be public intellectuals, those with visibility in the political mainstream. Scientists hope that the rest of the biosciences, as well as the broader society, will catch up to them in their understanding of health equity. They maintain that they hold a special view on the most essential piece of the puzzle: gene-environment relations.

In this style of thought, public outreach and education are top priorities. In fact, scientists write their commitments into their funding proposals and seek evidence of social goals in their reviews of others.[64] Scientists engage community groups and community-based representatives on ethical matters in the design phase of project planning.[65] Funding agencies sponsor science advisory panels to broker research proposals between postgenomic research teams and the public.[66]

Interdisciplinary collaboration is another leading priority. Collaborations in the form of seminars, conferences, and joint publications allow scientists to bridge the social-natural science chasm, while providing opportunities for geneticists to gain expertise on the environmental context. Scientists become well versed in talking about their own biases and assumptions and the potential for their identities to affect minority recruitment. Owning up to one's partiality, if not subjectivity, prior to launching projects is a mainstay of the postgenomic style of thought.

Finally, the postgenomic style of thought assumes the importance of local empowerment through knowledge production. Scientific collectives in the United Kingdom, United States, Canada, and Australia push models such as collective innovation, genomic sovereignty, and multinational partnership in the interest of offering training and supportive outsourcing to the developing world.[67] Individual researchers trade on their knowledge of specific communities to shuttle resources to underserved and vulnerable populations.[68]

This is not so different from the charisma-driven style of thought by which Steven Shapin has characterized the biotechnology industry in the latter half of the twentieth century.[69] Principal investigators and project leaders overtly rely on personality and display personal enthusiasm for social agendas to advance their scientific interests. Yet, it is a step beyond prior styles of thought that allowed scientists to maintain a secondary though supportive role in the mainstream political causes of the day or those for which scientists have assumed "moral equivalence" with the general public on political matters.[70] Today's style of thought allows geneticists to openly move through the biosciences and align with various nongenomic scientists in the effort to bring a social equality framework into the heart of research. It also permits postgenomic scientists to make their research goals take priority over health disparities tactics that take a more sociological approach.

The State of the Science

Despite the generation of a wealth of data, results about the causes and consequences of health disparities have been inconclusive. Studies on anthropometric traits associated with health disparities have come up with no replicable variants of genome-wide significance for common measures.[71] Meanwhile, replications have been made in multiracial samples that included racial groups that exhibit disease disparities.[72] Researchers continue to call for novel postgenomic models for understanding populations of shared genetic ancestry that exhibit different disease prevalences.[73] Researchers continue to collect comparative data. However, to date many have not produced results. The Human Heredity and Health in Africa Project and Southern African Human Genome Programme, for example, have only just begun releasing funds.[74] Pharmacogenomic companies are leading the way in whole-genome sequencing of non-European genomes,[75] putting a market-driven spin on the creation of therapies for minority populations.

Though postgenomic science is increasingly dedicated to the collection of environmental factors, its methodology continues to be gene hunting. Project leaders continue to advocate for "diversifying ethnic representation," but within a next-generation genomic framework.[76] Scientists most commonly seek copy number variants, epigenomic regions associated with communicable diseases, and rare variants. As it stands, there is a hierarchy within the interdisciplinary enterprise that postgenomics rests on based on how the public values sciences. Just as genomic scientists privilege genetic explanations in the last instance, the broader society currently puts a higher price on natural scientific solutions to social justice dilemmas. The genome remains the final explanandum for human quandaries.

This has important implications for the success of the postgenomic fight for health justice. Efforts to apply genomics to health disparities have not produced the benefits that scientists promised, because the social context has not been adequately addressed. Health equity is addressed at the level of the research project, as genetics becomes the leading way to address issues such as racism in health outcomes. Because scientists define the environment in terms of individual characteristics and health behaviors, such as body mass index and dietary habits, socioeconomic structures and the built environment are left poorly characterized. Social issues far afield from the immediate biosciences, such as institutionalized racisms and neighborhood effects, are not considered in this paradigm (see chap. 10 of this volume).

From Bench to Trench

This chapter has shown that, in the past decade, frontiers in genetic science and public health have converged to make health justice a postgenomic catchphrase and postgenomics a leading health justice tool. Examining the relationship between health justice and postgenomics from the angle of biopolitics and the sociology of knowledge demonstrates the interdependence of research and governance. The present form of postgenomics is not the result of an inert or inevitable set of technological advances.

Rather, it is shaped by the biopolitical architecture of governance dedicated to managing the most fundamental aspects of life itself.[77]

As genetic sciences move from bench to bedside, or, as researchers more aptly suggest, to the trenches of public health, social justice will increasingly be rendered a biomedical problem, displacing social scientific research as the main branch of disparities research. In this biopolitics of postgenomics and health justice, high-profile policy makers and scientists supersede social scientists and humanists as the trustees of egalitarianism. Research inclusion, once a nonissue for genetic sciences, becomes the justification for enlarging bioscience budgets. The "truth" of gender, race, and the social environment increasingly bears out in the domain of bioscience, potentially obfuscating the role of social structural violence and inequality in social conditions.

Postgenomic researchers have every incentive to keep genomics as the primary focus of study. As Pierre Bourdieu argues in his discussion of fields and capitals, sciences hold a unique status in society.[78] Society attributes a special kind of reason to sciences and credits sciences with describing reality. For geneticists, their capital, or cache, with the public is their ability to define biological reality. Continuing to monopolize that definition is what keeps the field funded by the billions.

Of course, individual researchers have different personal connections to the work they do.[79] But their collective initiation into the field, as well as their need to compete for scarce resources, keeps researchers acting in the interests of the field.[80] There are also broader structural reasons for maintaining a predominantly biological focus in matters of health justice. U.S. institutions spend billions more on life science research than social sciences and humanities combined. For example, in 2009, universities spent nearly $33 billion on biosciences and only $2 billion on social scientific research. The National Science Foundation spent $3.7 billion on science and engineering and $200 million on social science research. Health sciences institutions such as Johns Hopkins University were funded on the order of 26 times greater than leading social science institutions. Likewise, Health and Human Services was the greatest funder of university research.[81]

Interdisciplinary health disparities research affords postgenomic life scientists an expanded professional jurisdiction and greater symbolic reach.[82] Managing controversies associated with race and defining the role of social environments in health also help contemporary scientists clear their field's reputation regarding prior racisms such as eugenics (see chap. 8 of this volume). Connecting with the larger debate on race—whether disputing solutions, providing new visions of taxonomy, or permitting the influx of racial classification by self-identification—scientists expand their public image and social relevance. Thus, for health justice to be more than a watchword for obtaining funding, the biopolitics of emergent science must be understood.

Notes

1. Foucault, *Essential Foucault*, 202.
2. Foucault, *The History of Sexuality*, 143.
3. On the rise of "omics," see chap. 1 of this volume.
4. Foucault, *History of Sexuality*.
5. This analysis is taken from a review of Health and Human Services research priorities listed in its past five annual budgets.
6. Epstein, *Inclusion*, 17.
7. P. Hall, "Policy Paradigms, Social Learning, and the State," 279.
8. Pohlhaus and Cook-Deegan, "Genomics Research."
9. For analysis of the U.K. national biobank and population genomics in the United Kingdom, see Tutton, "Biobanks and the Inclusion of Racial/Ethnic Minorities"; and Smart et al., "Can Science Alone Improve." For a review of population biobank literature, see Corrigan and Tutton, "Biobanks and the Challenges of Governance."
10. Petryna, *When Experiments Travel*.
11. Whitmarsh, *Biomedical Ambiguity*.
12. Duster, "Race and Reification in Science," 1050.
13. N. Rose, *The Politics of Life Itself*, 12.
14. Bliss, "Translating Racial Genomics"; Rabinow and Rose, "Biopower Today."
15. Epstein, "Bodily Differences and Collective Identities"; Nelson, *Body and Soul*.
16. Epstein, *Inclusion*.
17. Carter-Pokras and Baquet, "What Is a 'Health Disparity'?"
18. Bliss, *Race Decoded*.
19. Epstein, *Inclusion*.
20. National Institutes of Health, "Understanding Our Genetic Inheritance," and "NIH Guide."
21. Collins and Galas, "New Five-Year Plan."
22. Bliss, "Genome Sampling and the Biopolitics of Race."
23. Stoto, Behrens, and Rosemont, "Healthy People 2000." Federal agencies attempted to implement Directive No. 15, a policy that required all publicly funded government agencies to use a racial classification system of White, Black, American Indian and Alaska Native, and Asian or Pacific Islander to report inclusion and participation in programs; see U.S. Office of Management and Budget, "OMB Directive No. 15." On the institutionalization of Directive No. 15, see Bliss, "Genome Sampling and the Biopolitics of Race."
24. National Institutes of Health, *NIH Revitalization Act of 1993*.
25. Center for Disease Control and Prevention, "Use of Race and Ethnicity."

26. Bliss, *Race Decoded*; Reardon, *Race to the Finish*.

27. Bowcock and Cavalli-Sforza, "Study of Variation in the Human Genome"; L. Roberts, "How to Sample the World's Genetic Diversity."

28. Reardon, *Race to the Finish*.

29. UNESCO, "Human Genome Diversity Project."

30. Knight, "Gene Project Deemed Unethical"; UNESCO, "Bioethics and Human Population Genetics Research."

31. U.S. Department of Energy (DOE) and the Human Genome Project, "To Know Ourselves."

32. Collins, Brooks, and Chakravarti, "DNA Polymorphism Discovery Resource."

33. Ibid., 688.

34. Bliss, *Race Decoded*.

35. Cf. Craig Venter in Buerkle, "'Wondrous Map' of Gene Data."

36. Weiss, "Genome Project Completed."

37. Xie et al., "Molecular Basis of Ethnic Differences."

38. Kahn, "Beyond BiDil."

39. Bliss, *Race Decoded*; Murray, "Race, Ethnicity, and Science."

40. Wade, "Genome Mappers Navigate the Tricky Terrain of Race."

41. Chandler, "Heredity Study Eyes European Origins"; Phillips et al., "Potential Role of Pharmacogenomics."

42. National Institutes of Health, "Report of the First Community Consultation."

43. U.S. Department of Health and Human Services, "Tracking Healthy People 2010."

44. U.S. Department of Health and Human Services, "Strategic Research Plan and Budget."

45. Risch et al., "Categorization of Humans in Biomedical Research," 1.

46. Wade, "Race Is Seen as Real Guide."

47. Burchard et al., "Importance of Race and Ethnic Background," 1174.

48. Wade, "Toward the First Racial Medicine."

49. Henig, "Genome in Black and White (and Gray)."

50. Ibid.

51. U.S. Department of Health and Human Services, "FY 2012 President's Budget for HHS," and "FY 2013 President's Budget for HHS."

52. U.S. Department of Health and Human Services, "HHS Strategic Plan, 2007–2012."

53. National Institutes of Health, "RFA-CA-09-001."

54. National Human Genome Research Institute, "National Human Genome Research Institute NIH Health Disparities Strategic Plan."

55. Center for Disease Control and Prevention, "Public Health Genomics."

56. National Institutes of Health, "Common Fund Strategic Planning Social Media Summary."

57. National Institutes of Health, "NIH FY 2012 Online Performance Appendix."

58. National Institutes of Health, "Genes, Environment and Health Initiative."

59. GENEVA, "GENEVA Study Overview"; U.S. Department of Health and Human Services, "FY 2013 President's Budget for HHS"; National Institutes of Health, "Common Fund Strategic Planning Social Media Summary."

60. National Institutes of Health, "Global Health Meetings."

61. Francis Collins, interview with author, April 2007.

62. Charles Rotimi, interview with author, June 2007.

63. David Altschuler, interview with author, April 2007.

64. Bliss, *Race Decoded*.

65. De Vries, Slabbert, and Pepper, "Ethical, Legal and Social Issues."

66. Jasanoff, *Reframing Rights*.

67. On collective innovation, see Ozdemir et al., "Towards an Ecology of Collective Innovation." On genomic sovereignty, see Benjamin, "Lab of Their Own." On multinational partnership, see, e.g., the following press releases produced for the Broad Institute of MIT and Harvard: N. Davis, "Mexico-US Collaboration Launched"; Cooney, "Two-Way Exchange."

68. Bliss, "Marketization of Identity Politics."

69. Shapin, *Scientific Life*.

70. K. Moore, *Disrupting Science*; Shapin, *Scientific Life*.

71. See, e.g., Kang et al., "Genome-Wide Association of Anthropometric Traits"; Mathias et al., "Genome-Wide Association Study."

72. See Hutter et al., "Replication of Breast Cancer GWAS Susceptibility Loci"; Ruiz-Narváez and Rosenberg, "Validation of a Small Set of Ancestral Informative Markers."

73. Rotimi, "Health Disparities in the Genomic Era"; Rotimi and Jorde, "Ancestry and Disease in the Age of Genomic Medicine."

74. Gross, "African Genomes."

75. See, e.g., Schuster et al., "Complete Khoisan and Bantu Genomes from Southern Africa."

76. Rotimi, "Health Disparities in the Genomic Era"; see also Bustamante, De La Vega, and Burchard, "Genomics for the World"; and Kittles, "Genes and Environments."

77. N. Rose, *The Politics of Life Itself*.

78. Bourdieu, *Pascalian Meditations*.

79. Bliss, *Race Decoded*.

80. Bourdieu, *Science of Science and Reflexivity*.

81. Bliss, *Race Decoded*.

82. Bliss, "Translating Racial Genomics."

10

The Missing Piece of the Puzzle?

MEASURING THE ENVIRONMENT IN THE POSTGENOMIC MOMENT

Sara Shostak and Margot Moinester

In the years leading up to the mapping of the human genome, its advocates promised that the Human Genome Project (HGP) would unlock the secrets of human health, illness, and identity. In 1988, a U.S. National Research Council Report argued that sequencing the human genome would "allow rapid progress to occur in the diagnosis and ultimate control of many human diseases."[1] In his tireless advocacy for the HGP, Francis Collins, then director of the National Human Genome Research Institute (NHGRI), proclaimed that "it is hard to imagine that genomic science will not soon reveal the mysteries of hereditary factors in heart disease, cancer, diabetes, mental illness, and a host of other conditions."[2] James Watson, the first director of the HGP, pronounced, "We used to think our fate was in the stars. Now we know, in large measure, our fate is in our genes."[3]

More than a decade into the postgenomic era, the improvements to individual and public health that serve as the primary rationale for massive public investiture in human genomics research have yet to be realized.[4] Researchers in both the social sciences and life sciences have argued that one of the paradoxes of the HGP is that it highlights the role of the environment as a "major piece of the puzzle" of human health and illness.[5] As Sarah S. Richardson and Hallam Stevens note (see chap. 12 of this volume), "Once-marginal critiques of a gene-centric vision of the life sciences are now moving to the center of the action." Indeed, a defining focus of postgenomic research has become elucidating the mechanisms and effects of "gene-environment interaction": "For the promise of genomics research to be fulfilled, we must move beyond the discovery of genetic variants associated with disease to understanding *how chemical changes to genes affect cell function in the context of complex environmental exposures*."[6]

Early advocacy for the concept of gene-environment interaction emerged from the environmental health sciences.[7] For nearly a decade, environmental health researchers have argued that research on gene-environment interaction is a necessary corrective to the HGP's "genocentric" view of human health and illness: "genocentric views reflect a fundamental misunderstanding of the disease process, and have led to unrealistic expectations and disappointment."[8] Specifically, they contend that research on gene-environment interaction is the key to explaining when and how genetic variations shape human health and illness: "Differences in our genetic makeup certainly influence our risks of developing various illnesses. . . . We only have to look at family medical histories to know that is true. *But whether a genetic predisposition actually makes a person sick depends on the interaction between genes and the environment*."[9]

In the postgenomic era, claims about the explanatory power of gene-environment interaction have gained traction also among genomic scientists (see chap. 2 of this volume). High-profile genomics researchers now aver that the "huge amounts of exuberance and enthusiasm" about the completion of the HGP led to the "overstating" of the immediate promise of genomics.[10] In myriad accounts of the relatively limited power of genomic research to explain variance in disease risk, leading genomic scientists argue that understanding gene-environment interaction will be an essential part of "translating genome based knowledge into health benefits."[11] Similarly, researchers have suggested that better measurements of environmental exposures may clarify associations between genetic variations and diseases that "currently seem completely random":[12] "Failure to account for [the] . . . additional level of complexity associated with environmental exposure and the resulting changes in gene expression may explain why most of the variance in disease risk has not been explained by genetic analysis alone."[13] Similarly, some scientists point to the complexity of the environment—and the inadequacy of measures thereof—in comments about the ongoing challenge of replicating results from genomic studies, including the relatively weak results generated by many genome-wide association studies (GWAS).[14] The leadership of the NHGRI now advocates for the "integration" of genomic information with data on "environmental exposure" in order to generate "a much fuller understanding of disease aetiology."[15]

Focusing on the environment also has offered biomedical researchers a means of linking genomics research with pressing public concerns regarding health disparities in the United States (see chap. 9 of this volume). Many scientists—and scientific institutions—have suggested that genetics has a role to play in reducing health disparities; in 2003, the NHGRI named the need to develop "genome-based tools" to "address health disparities" as one of its "grand challenges."[16] However, the prediction that genetics would offer a powerful way to address or alleviate health disparities has not been borne out, even in cases in which genetic polymorphisms linked to disease have been consistently identified.[17] Consequently, in recent publications, leading genomics scientists have argued that integrating environmental measures into biomedical research will be critical to efforts to remediate pervasive inequalities in health status across populations.[18] Epigenetics, in particular, promises new ways of understanding how racial and socioeconomic differences—as manifest in differences in sociomaterial environments—become embedded biologically.[19]

Consequently, a hallmark of the postgenomic era is the imperative that scientists elucidate the role of the environment in shaping the processes and outcomes of gene action. One claim is that research on gene-environment interaction will contribute to efforts to understand and ameliorate health disparities in the United States: "a more nuanced understanding of how genes interact with each other and with environments is necessary to fully understand if and how genetic variation contributed to health disparities."[20] A related claim is that research on gene-environment interaction will elucidate the role of social and cultural factors in health outcomes: "genetic information may point toward better identification of which kinds of social and cultural factors matter, why they matter, and when they matter for public health."[21] These imperatives are manifest in the proliferation of research in the life sciences that takes gene-environment interaction as its focus,[22] as well as increasing interest in gene-environment interaction among social scientists.[23]

These efforts face numerous challenges. To begin, scientists conceptualize and operationalize "the environment" in their research in tremendously varied ways. Even within the life sciences, the environment may refer to the cell (the environment of the gene), endogenous hormonal profiles (the environment of the cells), indoor or outdoor ambient environments (the environment of the human body), social networks, poverty, and/or stressful life situations (the social environment); individual behaviors, such as diet and exercise, may also be included in definitions of "the environment." At the same time, there is increasing heterogeneity in the temporal horizon of exposures of interest, with prenatal exposures, chronic low-dose exposures, and life-course perspectives taking their place alongside long-standing concerns about the immediate effects of acute toxic exposures.[24] Dramatizing the difference between lab-based approaches and social science perspectives, recent publications by sociologists argue that the actual lived environments that shape population health are not merely "little Petri dishes" but rather include dynamic processes unfolding across time and place.[25] So, although life scientists and social scientists agree that "environmental context matters,"[26] conceptualizing and operationalizing environmental contexts in postgenomic research and elucidating their contributions to population health will be no small feat.

Our goal in this analysis is to highlight how within fields of knowledge production specific techniques of measuring the environment create "regimes of perceptibility" in which particular aspects of the environment become more and less visible, appear as material objects, and populate the worlds of the lab, the clinic, and the community.[27] Regimes of perceptibility are produced within disciplinary or epistemological traditions that regularize, standardize, and sediment the contours of perception and imperception.[28]

Therefore, in this chapter we extend the volume's discussion of the contested and unsettled nature of the postgenomic age by comparing recent efforts to meet the challenge of measuring and conceptualizing the environment in two very different epistemological traditions.[29] First, we examine exposomics, an emergent postgenomic field that seeks to assess "the whole environment we have inside our bodies."[30] We selected exposomics as a focal case from among "omics" approaches to gene-environment interaction in the life sciences for three reasons: its primary goal is to measure environmental exposures; having emerged in 2005, it is chronologically a postgenomic science; and, although exposomics is still a field in formation, it has generated enough literature for a meaningful examination of its key concepts and strategies.[31] We contrast the environment as made visible in exposomics with measurements of the environment in contemporary social epidemiological and sociological research on neighborhood effects on health. We chose to focus on the neighborhood effects literature as a means of highlighting the emphasis, within the social sciences, on structures and social processes that are not easily captured in individual-level studies and that research has identified as relevant to understanding the production and maintenance of health disparities in the United States.[32] Indeed, health disparities researchers argue that "your zip code is more important than your genetic code,"[33] and neighborhood effects research has expanded significantly in the postgenomic era. This comparison elucidates the very different conceptualizations and techniques of measuring the environment that characterize the postgenomic moment; it also suggests that the tensions between different disciplinary approaches to the environment—and whether or how they are resolved—are a central dynamic in the development of postgenomic knowledge production.

Environment as Exposome: Looking inside the Human Body

The term "exposome" was coined by Christopher Wild, a molecular epidemiologist whose career has focused on the "interplay" between environmental and genetic risk factors.[34] According to Wild, the motivation for developing the concept was "to draw attention to the critical need for more complete environmental exposure assessment in epidemiological studies."[35] The goal of exposome research is to characterize "life course environmental exposures (including lifestyle factors) from the prenatal period onward."[36]

Exposome research is being developed primarily by environmental health scientists, and especially by molecular epidemiologists. Broadly, it is part of "exposure science," an effort to leverage new forms of science and technology as techniques both to identify sources of environmental exposures and to account for their consequences in living systems.[37] Advocates of "exposomics" contend that environmental epidemiology has been limited by a "parochial" division of exposures into categories such as air and water pollution, occupation, diet and physical activity, and stress.[38] In contrast, they propose a reconceptualization of environmental exposures that encompasses a wide variety of external and internal factors relevant to human biological processes and their health effects. The exposome consists of every exposure that an individual experiences, from conception to death.[39]

Publications on the exposome emphasize the importance of the environment, defined very broadly as "nongenetic" factors,[40] to human health: "Investigations of human twins and results from more than 400 genome-wide association studies (GWAS) indicate that genetic factors contribute about 10–30% of the risks of cancer and cardio-vascular diseases. . . . *Thus, it appears that non-genetic factors (i.e. 'the environment') are the major causes of chronic diseases in human populations.*"[41] Recent papers emphasize techniques for doing "Exposome Wide Association Studies" as a means of discovering "components of the exposome" that cause chronic disease.[42] Proponents of exposomics argue that it represents an "unbiased approach to discovering the causes of disease."[43] However, there is significant variation in the kinds of exposures included in different approaches to exposome research.

The broadest definition of the exposome includes *internal exposures* (e.g., processes internal to the body such as metabolism, endogenous circulating hormones, body morphology, physical activity, gut microflora, inflammation, lipid peroxidation, oxidative stress, and aging), *specific external exposures* (e.g., radiation, infectious agents, chemical contaminants and environmental pollutants, diet, lifestyle factors [e.g., tobacco, alcohol], occupation and medical interventions), and *general external exposures*, defined as wider social, economic, and psychological influences on the individual (e.g., social capital, education, financial status, psychological and mental stress, urban/rural environment, and climate).[44] More narrow definitions of the exposome emphasize the internal consequences of individual behaviors, discounting factors from the broader domains described above: "the things that people generally associate with the environment, namely, air and water pollution, hygiene and sanitation, smoke from fuel combustion, and occupation, collectively contribute only about 7–10% to the burden of chronic diseases. . . . Rather, *the major environmental risk factors identified thus far are smoking, overweight, and the diet*."[45] To date, however, the best evidence for the exposome concept comes from molecular measures of specific external exposures, such as occupational exposure to benzene.[46] Approaches to exposomics that include the social environment require a more extensive set of measures that include multiple levels of analysis.[47] Critics of this approach have highlighted the challenge of combining in one study "everything from sociological stresses to psychological stresses to chemical, physical, and biological stresses" and have suggested the importance—and complexity—of "a common metric" that can be used to make meaningful comparisons across these categories.[48]

Already, there is a clear tendency to look for this "common metric" inside the human body. Exposome research focuses on "the body's *internal chemical environment*." In this approach, exposures, of whatever sort, are measured in terms of "the amounts of biologically active chemicals *in this internal environment*."[49] Consequently, under both the more expansive and the more narrow definition of "exposures," the environment in exposome research is defined as a characteristic of individuals and measured as a discrete biomarker, or set of biomarkers, inside the human body.

Indeed, scientists articulate the exposome as the "complement" to the genome:[50] "Whereas the genome gives rise to a programmed set of molecules . . . in the blood, the exposome is functionally represented by *the complementary set of chemicals derived from sources outside of genetic control.*"[51] Advocates of exposomics argue that "we need to be able to analyze 'the environment' in a way analogous to the genetics' community analysis of variation,"[52] rather than "relying on questionnaires to characterize 'environmental exposures.' "[53] They call for a "Human Exposome Project" similar in scope to the HGP.[54] Indeed, they contend that "inaccurate and imprecise environmental data lead[ing] to biased inferences" pose a threat to current efforts to assess gene-environment interaction.[55]

Thus, even as it greatly expands the kinds and timing of exposures that genome scientists deem relevant to human health, exposomics simultaneously materializes these exposures as individual attributes. By focusing wholly on the internal environment—which inevitably will be unique to each individual—exposomics promises better assessment of individual risks. As such, it seems well suited to clinical interventions for individuals. As one researcher imagines, "These are things that you could eventually see in a doctor's office. . . . It wouldn't take too great a stretch of the imagination to expect physicians to regularly screen patients for 100 or so of the most implicated chemicals to assess disease risk." In this way, environmental health and illness are aligned with "the vision of the future of medicine being more predictive and more personalized."[56]

Whether and how exposomics can elucidate our understanding of community-level exposures and contribute to disease prevention at a population level remains unclear. Certainly, there is a possibility that individuals who learn about the contents of their "exposome" may be motivated to take steps to reduce their exposures, whether through individual behavior change, community organizing, or efforts to change state and federal environmental policy.[57] However, as we detail in the following pages, exposome research may have the unintended consequence of rendering invisible those aspects of the environment that are not easily reduced to individual-level measurements. This possibility becomes clear when we examine a contrasting scholarly literature in which the environment is conceptualized at the neighborhood level.

Environment as Ecology: Studying Neighborhoods

Proponents of ecological analysis in health research contend that studying individuals divorced from the broader context that conditions their lives leads to an incomplete understanding of disease etiology. As a leading social epidemiologist asserts, "We live embodied: 'genes' do not interact with exogenous (that is, outside of the body) environments—only organisms do, with consequences for gene regulation and expression."[58] Ecological research directly attends to this exogenous environment and underscores how "pairings, families, peer groups, schools, communities, cultures and laws are all contexts that alter outcomes in ways not explicable by studies that focus solely on individuals."[59] The underlying notion is that macro-level variables and social processes (e.g., selection, distribution, interaction, adaptation, and other responses) can independently affect individual outcomes or modify how individual characteristics are related to these outcomes.[60] Nicholas Christakis and James H. Fowler, for example, highlight the social nature of obesity by demonstrating that obesity spreads in social networks in distinct, quantifiable patterns that depend on the nature of social ties.[61] In contrast to the exposomics approach, ecological researchers contend that the study of group-level variables has a distinct analytic value. As such, they attempt to make visible aspects of the environment that cannot be captured at the individual level and yet affect both individual and population outcomes.

Within the ecological tradition, *place* has long been viewed as an important context for the study of a wide range of individual and population outcomes, including health. Scholars argue that places matter because they embody social significance through the interaction of material form, geography, and inscribed meanings and can affect people's health behaviors and outcomes.[62] Neighborhoods play a particularly prominent role within this research. As geographically bound, socially defined entities, neighborhoods possess physical and social attributes that may shape residents' environmental exposures and construct or constrain opportunities for various behaviors and actions through a complex system of interactional processes.[63] Moreover, because place of residence is strongly patterned by social position and race/ethnicity, neighborhoods provide an important context for the study of disparities and "differential vulnerability."[64] Neighborhoods have also been shown to possess symbolic significance, with implications for inequality: "Neighborhoods have reputations—people act as if neighborhoods matter and thus imbue neighborhoods with meanings that have profound importance in the social reproduction of inequality by place."[65]

The operationalization of "neighborhoods" in health research has evolved over the past century. The systematic study of neighborhoods first became prominent in the beginning of the twentieth century when urban sociologists of the classical Chicago School employed a dialectic process of systematic group-level observations and theory building to elucidate important determinants of health that operate at the neighborhood level.

These scholars found that neighborhoods characterized by low economic status, ethnic and linguistic heterogeneity, and residential instability experienced disproportionately high rates of crime, infant mortality, low birth weight, tuberculosis, physical abuse, and other health-related outcomes.[66] They also found that these outcomes persisted over time despite residential turnover, which led them to argue that neighborhoods possess "relatively enduring features and emergent properties that transcend the idiosyncratic characteristics of particular ethnic groups that inhabit them."[67] Similarly, contemporary multilevel analysis has found a generally strong association between living in an area of concentrated deprivation and a wide range of health outcomes, including type 2 diabetes, low birth weight, mental health, cardiovascular disease, and general mortality.[68] These associations both contextualize and transcend individual-level attributes and have led scholars to view the local environment as "*an independent causal agent* in producing poor health."[69]

Over the past two decades, research on neighborhoods and health has grown significantly in epidemiology and the social sciences. This increase is due in part to a growing sense that individual-based explanations of the causes of poor health are insufficient and also to heightened public concerns regarding health disparities.[70] As such, recent scholarship attends both to the effects of neighborhoods on individual outcomes and to explanations for variations in health outcomes across neighborhoods. This research focuses on features of the neighborhood physical and social environment. The physical environment includes traditional environmental exposures (e.g., air pollution and exposure to toxic substances), as well as aspects of the built environment (e.g., land use, transportation, street design, public spaces, and access to resources such as recreational opportunities and healthy foods). The social environment includes factors such as psychosocial stress, the degree of social cohesion between neighbors, levels of crime and violence, the presence of social norms, and features of the social organization of places.[71] Specifically, this research investigates the multiple pathways through which various aspects of a neighborhood's physical and social environments may get "under the skin," as well as the structural forces driving ecological differentiation in health outcomes.

Neighborhood effects scholars also emphasize the "need to consider not only how local areas or neighborhoods affect health but also how the broader spatial context within which neighborhoods are situated may add to or modify the effect of local areas."[72] The study of neighborhood effects is a subset of the broader ecological analysis of spatially defined contexts on health.[73] Proponents of extra-neighborhood spatial analysis contend that significant advancements in geospatial and statistical methods, data, and technologies hold the promise of helping researchers to more accurately characterize exposures to different social and physical environments potentially important for health behaviors and outcomes. As a team of researchers argued, "charting an individual's personal geography through multiple 'places' and 'contexts' over the day, week, month or even the life-course, will give us improved measures of exposure and allow us not only to understand which environments are most salient for health in terms of location and duration but also how an individual's personal characteristics mediate this relationship."[74]

In directly attending to neighborhood-level variability in environmental exposures and health behaviors and outcomes, ecological research is uniquely positioned to elucidate the historically contingent, spatial and temporal processes driving health disparities more broadly. In the context of the United States, the weight of evidence suggests that American cities are significantly differentiated along socioeconomic and racial lines,[75] and that this differentiation in turn corresponds to the spatial differentiation of neighborhoods along many dimensions of health.[76] This variance bears directly on health disparities as consistently "population patterns of good and bad health mirror population distributions of deprivation and privilege."[77] Moreover, despite significant changes in the structure of American cities over the past half century, ecological scholars argue that the *relative position* of a neighborhood in terms of its economic, political, and social characteristics remains "remarkably stable" over time.[78] This stability has significant implications for the durability of health disparities at the neighborhood level. These findings have subsequently led researchers and policy makers to "now accept the role of neighborhood social and material context in producing and maintaining inequalities in health and life chances."[79]

By focusing on the larger environmental context, ecological research attempts to make visible social institutions and processes that bear on individual and population health outcomes in ways not explicable in studies that focus solely on individuals or their internal chemical environment. Research on neighborhoods, specifically, may help to capture important characteristics of the social and physical environment and elucidate the mechanisms linking these characteristics to health outcomes. In turn, this research may help to better illustrate "how we, like any living organism, literally incorporate, biologically, the world in which we live, including our societal and ecological circumstances."[80] Additionally, this line of inquiry holds promise for revealing the ecological distribution of environmental factors important in gene regulation and expression. As such, it would help to expose "differential vulnerability" across space, which in turn may inform disease prevention at a population level and efforts to reduce health disparities.

The Challenge of the Postgenomic Environment

Our analysis both points to and recapitulates a central tension of the postgenomic moment. Postgenomics is characterized by an emerging scientific consensus that "the environment" matters, a broad commitment to elucidating how gene action and expression are shaped by particular environmental contexts, and an aspiration that research on gene-environment interaction will contribute to efforts to ameliorate health disparities in the United States. However, there is no consensus about what is meant by "the environment" or how it should be measured. As such, even as life scientists and social scientists express a shared enthusiasm for research on "gene-environment interaction," what this means, in practice, varies dramatically across disciplines. Our analysis highlights particularly the difference between exposome research, which conceptualizes the environment as an individual-level, internal trait—very much like a genome—and neighborhood-level research, which is defined by its focus on the social structures and social processes that shape population health. We conceptualize this difference as a consequence of extant regimes of perceptibility within epistemological traditions. More broadly, however, these differences point to the degree to which defining the environment is a central challenge in postgenomic knowledge production.

Exposure scientists contend that postgenomic conceptualizations of the environment such as the "envirome" or "exposome" highlight its importance and, indeed, its "complementarity" with the genome.[81] Insofar as genomics serves as the model, source of technology, and historical inspiration for exposomics, it is unsurprising that exposomics inclines also toward biomedical conceptualizations of the environment that focus on individual-level rather than population risks. We expect that exposomics will increase our understanding of life-course exposures and their consequences for individuals. We tend to agree with the advocates of exposomics that better measurements of the environment are critical to understanding gene-environment interactions. However, just as "the reason to study prostate cancer genetics is not to address health disparities . . . it is to determine the genetics of prostate cancer,"[82] the reason to study the exposome is not to address health disparities but to assess environmental exposures across the life course. To be fair, advocates of exposomics do not position themselves as health disparities researchers. Rather, as described above, one rationale for exposomics is that it will contribute to our understanding of gene-environment interaction, which, in the postgenomic moment, is offered by many scientists as central to efforts to understand and ameliorate health disparities.

At the same time, -omic conceptualizations of the environment risk obfuscating the full range of environments that are relevant to population health. Insofar as they molecularize and individualize "the environment," these techniques impose a reductionist regime of perceptibility in which social structure and social processes, especially, tend to disappear. This raises the concern that, particularly in the context of pervasive cultural and political dynamics favoring individualistic explanations of population health and discounting social determinants of health disparities,[83] exposomics might have the unintended consequence of recasting even the most socially determined exposures as individual traits.

The case of diet highlights what is at stake in these different approaches to conceptualizing and materializing the environment. A recent paper in the *Annual Review of Nutrition* identifies the exposome as a "bold and visionary" conceptual development that highlights the role of diet "*as the greatest single source of chemical exposures*, including nutrients, nonnutritive chemicals, pesticides, and others."[84] Ecological analysis of neighborhoods also highlights the role of diet in shaping health outcomes, but rather than measuring diet within the internal chemical environment, this research examines the social institutions and processes that structure opportunities for access to different types of diets. Specifically, this line of research investigates "how the structure and organization of the neighborhood food environment (operationalized as the availability of grocery stores and fast-food restaurants) might influence food purchasing patterns and hence diet and diet-related chronic diseases."[85] This research variously conceptualizes the accessibility, availability, affordability, and quality of local food environments through a combination of geographic and quantitative analysis.[86] Furthermore, through employing qualitative techniques, recent research also investigates the social processes and symbolic relations between individuals and their food environment in order to examine the "dynamic relationships between meaning and action" regarding food purchasing behavior.[87] Collectively, this line of research conceptualizes "spatial variations in exposure to aspects of the local food environment as an underlying explanatory factor for social and spatial inequalities in diet and related health outcomes."[88]

Similar tensions are evident in postgenomic approaches to research on the health effects of stress. Exposomics scholars emphasize the importance of psychosocial stress in disease etiology.[89] Measuring stress within the internal chemical environment, however, obfuscates the conditions by which stress emerges. In contrast, neighborhood effects research directly attends to the social and physical context in which these conditions are embedded. As Rachel Morello-Frosch and Edmond D. Shenassa note, "Place-based stressors are biologically relevant components of the human environment and can function independently of individual-level stressors to determine health."[90] For example, research demonstrates a robust link between exposure to violent crime and exposure to environmental pollutants and stress.[91] The saliency of neighborhood context for psychosocial stress is further elucidated in the results of an experimental research study, which indicate that the move from a high-poverty neighborhood to a low-poverty neighborhood reduced residents' symptoms of distress and depression.[92] Collectively, this work makes visible the "geographies of exposure and susceptibility,"[93] whereas exposomic research reveals the biological pathways by which stress affects health outcomes.

Research on neighborhoods and health thus points to an alternative regime of perceiving the environment and its associations with health disparities. This perspective insists that health—of both individuals and populations—is an emergent property of systems "in which processes operating at the levels of individuals and populations are inextricably connected."[94] Therefore, addressing health disparities requires measures of those processes, not only inside the human body but also in the places where we live, work, and play. For example, by illustrating the ways that food consumption patterns vary spatially, neighborhood-level studies can provide insight into how individuals are differentially "exposed" as they navigate different environments over the life course. They also point to the irreducibly social nature of environmental exposures, which may be obfuscated by research that focuses only on the internal, molecular environment of individuals. This way of understanding the environment has some traction in genomics science. For example, the leadership of the NHGRI recently asserted that "most documented causes of health disparities are not genetic, but are due to poor living conditions and limited access to healthcare."[95] It remains to be seen whether or to what extent such insights will shape the NHGRI's postgenomic research agenda.

In conclusion—and to extend a commonly used metaphor—the postgenomic moment is characterized by an emerging scientific consensus that the environment is a "missing piece" of the "puzzle" of human health and illness. Nonetheless, there is no agreement regarding the size, shape, or location of this puzzle piece—or pieces. The concept of gene-environment interaction directs the search for the environment—and its effects on human health—toward the same place where one finds genes, that is, inside the body and at the molecular level. However, a variety of disciplines across the health and social sciences are proposing alternate conceptualizations and techniques for measuring the environment and its consequences for health, which may include but are not limited to its effects on gene expression. There are fundamental tensions between approaches to the environment that frame it as a molecular, individual trait—that is, as a version of the genome—and those that focus on multiple levels of analysis and techniques of measurement. The existence of these tensions is an essential aspect of postgenomic science. The process by which they are—or are not—resolved has the potential to redefine knowledge production in the postgenomic era.

Notes

1. Green et al., "Charting a Course for Genomic Medicine," 211.

2. Collins, "Contemplating the End of the Beginning," 643.

3. Quoted by Jaroff, "Gene Hunt."

4. See, e.g., Guttmacher and Collins, "Realizing the Promise of Genomics in Biomedical Research."

5. Arnaud, "Exposing the Exposome," 42; see also Pescosolido, "Of Pride and Prejudice."

6. Olden et al., "Discovering How Environmental Exposures Alter Genes," 834; emphasis added.

7. Shostak, *Exposed Science*, provides a detailed analysis of the environmental health sciences' engagement with genomics.

8. Olden and White, "Health-Related Disparities," 721.

9. National Institutes of Health, "Two NIH Initiatives Launch Intensive Efforts"; see also Schwartz and Collins, "Medicine."

10. Green, in Kaiser, "Genome Project."

11. Collins et al., "Vision for the Future of Genomics Research," 840; see also Guttmacher and Collins, "Welcome to the Genomic Era"; Schwartz and Collins, "Medicine."

12. Harmon, "Sequencing the 'Exposome.' "

13. Olden et al., "Discovering How Environmental Exposures Alter Genes," 834.

14. Patel, Bhattacharya, and Butte, "Environment-Wide Association Study"; MacArthur, "Why Do Genome-Wide Scans Fail?"

15. Green et al., "Charting a Course for Genomic Medicine," 208.

16. Sankar et al., "Genetic Research and Health Disparities," 2986. An early critique of efforts to link genetic research to health disparities argued that "the

question is not whether genetics has enhanced our understanding of the disease process in individuals. It clearly has. Rather, the question is whether, all things considered . . . genetics research is a particularly effective way to proceed with the effort to alleviate health disparities in the United States"; ibid., 2988.

17. Diez Roux, "Complex Systems Thinking and Current Impasses," 1628.

18. Green et al., "Charting a Course for Genomic Medicine."

19. Kuzawa and Sweet, "Epigenetics and the Embodiment of Race."

20. Diez Roux, "Complex Systems Thinking and Current Impasses," 1628; see also Olden et al., "Discovering How Environmental Exposures Alter Genes."

21. Bearman, "Genes Can Point to Environments"; see also chap. 2 of this volume.

22. Feil and Fraga, "Epigenetics and the Environment."

23. Pescosolido, "Of Pride and Prejudice"; Bearman, Martin, and Shostak, *Exploring Genetics and Social Structure*.

24. On prenatal exposures, see chap. 11 of this volume; on chronic low-dose exposures, see Altman et al., "Pollution Comes Home and Gets Personal"; on life-course perspectives, see Shanahan and Hofer, "Social Context in Gene-Environment Interactions."

25. Bearman, "Genes Can Point to Environments"; see also Boardman, Daw, and Freese, "Defining the Environment in Gene-Environment Research."

26. Olden et al., "Discovering How Environmental Exposures Alter Genes," 834.

27. Murphy, *Sick Building Syndrome and the Problem of Uncertainty*.

28. Ibid.

29. Our primary data on exposomics come from peer-reviewed journal articles, published from 2005 to 2012 ($n=24$); our sample includes reports of empirical research, review articles, and editorials/position statements. Our primary data on neighborhoods come from peer-reviewed journal articles and edited book volumes ($n=29$); similarly, this sample includes reports on empirical research, review articles, and editorial/position statements.

30. Harmon, "Sequencing the 'Exposome,'" 2.

31. To date, the concept of the exposome and associated measurement techniques have been taken up in research on diverse health outcomes, including adolescent idiopathic scoliosis (Burwell et al., "Adolescent Idiopathic Scoliosis"), cancer (Faisandier et al., "Occupational Exposome"; Smith et al., "Benzene, the Exposome and Future Investigations of Leukemia Etiology"), and metabolism (Ellis et al., "Metabolic Profiling Detects Early Effects").

32. Diez Roux and Mair, "Neighborhoods and Health"; Diez Roux, "Neighborhoods and Health"; Gee and Payne-Sturges, "Environmental Health Disparities"; Brulle and Pellow, "Environmental Justice."

33. For an example, see Marks, "Why your zipcode may be more important."

34. Wild is currently the director of the World Health Organization's International Agency for Research on Cancer (IARC). At the time he coined the term "exposome," he was professor of molecular epidemiology and head of the Center for

Epidemiology and Biostatistics at the University of Leeds. See http://www.iarc.fr /en/office-dir/director/dirbiography.php.

35. Wild, "Exposome," 24.

36. Wild, "Complementing the Genome with an 'Exposome,'" 1848.

37. National Academy of Sciences, *Exposure Science in the 21st Century.*

38. Rappaport and Smith, "Environment and Disease Risks."

39. Wild, "Exposome."

40. Ibid., 24.

41. Rappaport, "Implications of the Exposome for Exposure Science," 6; emphasis added.

42. Rappaport, "Biomarkers Intersect with the Exposome."

43. Rappaport et al., "Advances in Biological Monitoring," 83; emphasis added.

44. Wild, "Exposome."

45. Rappaport et al., "Advances in Biological Monitoring," 83; emphasis added.

46. Wild, "Exposome," 27.

47. Ibid.

48. Arnaud, "Exposing the Exposome."

49. Rappaport and Smith, "Environment and Disease Risks," 460; emphasis added.

50. Wild, "Complementing the Genome with an 'Exposome.'"

51. Rappaport, "Biomarkers Intersect with the Exposome," 2; emphasis added.

52. Balshaw, "Making the Case for Advancing the Exposome (or EWAS)."

53. Rappaport and Smith, "Environment and Disease Risks," 460.

54. Arnaud, "Exposing the Exposome." Further, they propose that the history of the HGP suggests that perceived barriers to technology development should not deter such efforts, as the sequencing technologies that resulted in the successful completion of the HGP did not exist at the time it was launched; Wild, "Complementing the Genome with an 'Exposome.'"

55. Harmon, "Sequencing the 'Exposome,'" 2.

56. Ibid.

57. Altman et al., "Pollution Comes Home and Gets Personal."

58. Krieger, "Embodiment," 351.

59. Susser, "Logic in Ecological," 825.

60. Diez Roux, "Bringing Context Back into Epidemiology"; Susser, "Logic in Ecological."

61. Christakis and Fowler, "Spread of Obesity."

62. Gieryn, "Space for Place in Sociology."

63. MacIntyre and Ellaway, "Neighbourhoods and Health."

64. Diez Roux and Mair, "Neighborhoods and Health"; Gee and Payne-Sturges, "Environmental Health Disparities."

65. Sampson, *Great American City,* 49.

66. Shaw and McKay, *Juvenile Delinquency and Urban Areas.*

67. Sampson, *Great American City,* 37.

68. Auchincloss et al., "Neighborhood Resources for Physical Activity"; Roberts, "Neighborhood Social Environments"; Leventhal and Brooks-Gunn, "Moving to Opportunity"; Latkin and Curry, "Stressful Neighborhoods and Depression"; Augustin et al., "Neighborhood Psychosocial Hazards and Cardiovascular Disease"; Diez Roux et al., "Neighborhood of Residence and Incidence of Coronary Heart Disease"; Yen and Kaplan, "Neighborhood Social Environment and Risk of Death."

69. Cummins, "Improving Population Health through Area-Based Social Interventions," 288; emphasis added.

70. Diez Roux and Christina Mair, "Neighborhoods and Health"; Diez Roux, "Neighborhoods and Health."

71. Diez Roux and Mair, "Neighborhoods and Health"; MacIntyre and Ellaway, "Neighbourhoods and Health."

72. Cummins et al., "Understanding and Representing 'Place' in Health Research," 1832.

73. Diez Roux and Mair, "Neighborhoods and Health," 134.

74. Cummins et al., "Understanding and Representing 'Place' in Health Research," 1835.

75. Sampson and Wilson, "Toward a Theory of Race"; Massey and Denton, American Apartheid.

76. Gee and Payne-Sturges, "Environmental Health Disparities"; Sampson, Morenoff, and Gannon-Rowley, "Assessing 'Neighborhood Effects.'"

77. Krieger, "Theories for Social Epidemiology in the 21st Century," 668.

78. Sampson and Morenoff, "Durable Inequality"; Sampson, Great American City.

79. Cummins et al., "Measuring Neighbourhood Social and Material Context," 249.

80. Krieger, "Embodiment," 351.

81. Patel et al., "Environment-Wide Association Study"; Wild, "Complementing the Genome with an 'Exposome.'" Scientists are not alone in their "omics" framing of the environment; the Environmental Working Group (EWG), a U.S.-based environmental advocacy group, has launched the Human Toxome Project (HTP) in order to define the "human toxome . . . the full scope of industrial pollution in humanity." See http://www.ewg.org/sites/humantoxome/about/.

82. Sankar et al., "Genetic Research and Health Disparities," 2988.

83. Krieger, "Stormy Weather," 2159.

84. Jones, Park, and Ziegler, "Nutritional Metabolomics," 186; emphasis added.

85. Thompson et al., "Understanding Interactions with the Food Environment," 116.

86. McKinnon et al., "Measures of the Food Environment."

87. Thompson et al., "Understanding Interactions with the Food Environment," 117.

88. Cummins, "Neighbourhood Food Environment and Diet," 196.

89. Wild, "Exposome," 24–25.

90. Morello-Frosch and Shenassa, "Environmental 'Riskscape' and Social Inequality," 1150.

91. Ellen, Mijanovich, and Dillman, "Neighborhood Effects on Health."

92. Leventhal and Brooks-Gunn, "Moving to Opportunity."

93. Jerrett and Finkelstein, "Geographies of Risk."

94. Diez Roux, "Complex Systems Thinking and Current Impasses," 1627.

95. Green et al., "Charting a Course for Genomic Medicine," 211.

11

Maternal Bodies in the Postgenomic Order

GENDER AND THE EXPLANATORY LANDSCAPE OF EPIGENETICS

Sarah S. Richardson

The neurologist and geneticist Michael Meaney argues that a stressed pregnant woman may produce offspring prone to anxiety, depression, schizophrenia, and suicide.[1] The psychiatrist Ray Blanchard warns that a mother with a large number of sons could damage subsequent sons in the womb owing to the accumulation of immune antibodies to male fetuses—causing later-born sons to become homosexual.[2] The recent popular science book *Origins: How the Nine Months before Birth Shape the Rest of Our Lives* (2010) reviews new research suggesting that a mother's stress level and dietary habits during gestation can "program" the fetus for a future of obesity, heart disease, and diabetes.[3]

Epigenetics, the study of how experiences, environments, and exposures alter gene expression, is a vibrant new area of postgenomic life sciences research. This chapter examines how maternal bodies are situated and valenced within this scientific field. Using texts and images from the scientific literature, as well as its public intellectual and popular reception, I document how epigenetics research situates the maternal body as a central site of epigenetic programming and transmission and as a significant locus of medical and public health intervention in the postgenomic age.

The science of maternal-fetal epigenetic programming converges with several major trends in twentieth- and twenty-first-century science, gender, and culture: from conceiving of motherhood as instinctual, selfless,

and intrinsically moral, to a cultural conception of mothering as an agential project of the self in which the mother's interests are often perceived to
be in tension with the child's;[4] from a psychosocial model of child development, to a model in which the critical factors in development are genetic
and neurological;[5] and from birth as the moment of personhood and medical concern, to conception, and even preconception, as the focal point of
political interest and biomedical intervention in reproduction.[6]

Epigenetic studies of "maternal effects" raise vital social, ethical, and
philosophical questions.[7] Is there a potential for this research to heighten
public health surveillance of and restrictions on pregnant women and
mothers through a molecular policing of their behavior?[8] How might this
new research participate in the often-troubled history of notions of the su-
preme role of the mother in normal and pathological development?[9] What
are the empirical and methodological implications of a research focus on
maternal effects to the exclusion of the larger social environment (and of
paternal effects)?[10]

This chapter touches on all of these important questions, albeit through
a side window. In the context of this volume, the central objective of this
chapter is to explore what epigenetic studies of maternal effects reveal
about the explanatory landscape of postgenomic science. Specifically, does
epigenetics represent a challenge to, or just another version of, genetic
determinism?[11] In this chapter I pose the figure of the maternal body in
epigenetics as an index case to examine central conceptual questions about
the definition, scope, and stakes of the "postgenomic" life sciences.

The figure of the maternal body as an "epigenetic vector," I argue, compels a different reading of the postgenomic commitment to complexity, to
anti-determinism, and to a biosocial conception of human heredity and
health than has generally been assumed. While scientists, social scientists,
and philosophers theorize epigenetics as a long-awaited turn toward holist
explanations of life and a vindication of their trenchant critiques of the
conceptual limits of genetic determinism and reductionism, in practice
the leading-edge research agendas, experimental designs, and therapeutic
interventions in human epigenetics complicate and may even foreclose
this vision. Research on maternal epigenetic programming points to an
emerging postgenomic explanatory order in which traditional forms of genetic determinism and reductionism are subtly reformulated. Rather than
conceptualizing genes as agents in a linear causal chain, epigenetics and
related postgenomics disciplines see genes as difference makers within
embodied contexts. This reformulation, made especially clarion by the

case of maternal bodies in epigenetic research, highlights the need for new analytic approaches to suit the forms of explanation increasingly prevalent in the postgenomic life sciences.

Epigenetics

Epigenetics is the study of molecular mechanisms that bring about a heritable or persistent change in gene function without changing gene sequence.[12] One such mechanism is DNA methylation, the process by which a methyl group (CH_3) is appended to the physical structure of the DNA molecule.[13] The presence of methylation at a particular gene locus typically prevents gene expression via physical obstruction of DNA transcriptase and other DNA-binding proteins.

Nonhuman animal studies allowing for experimental manipulation provide the strongest evidence for epigenetic mechanisms of gene expression modification. In mice with a gene variant linked to yellow fur color and obesity, a methyl-rich maternal diet during gestation epigenetically alters gene expression to yield brown offspring of typical body size.[14] In a species of vole, maternal melatonin reflecting the season epigenetically influences the fetus's coat thickness, preparing it for its future environment.[15] In rats, early maternal licking of pups epigenetically programs glucocorticoid receptor gene expression in the brain, resulting in a low-anxiety adult phenotype.[16] In all of these classic studies, the epigenetic modification is introduced via the behavior or physiology of the mother—a point to which I return below.

Epigenetic alterations of gene function may be fleeting or long-lasting. Typically, they are reversible. They can be passed through cell division from one cell to another. They can also be inherited from one generation to another through the gamete, cytoplasm, or reconstruction of the environmental cues activating the trait in each new generation. This is a form of indirect, rather than direct, inheritance.[17]

The extent and importance of early-life epigenetic programming in human health are not yet known.[18] A burgeoning scientific literature claims associations between prenatal hormonal, immunological, and nutritional exposures and adult phenotypes such as attention deficit/hyperactivity disorder, autism, schizophrenia, homosexuality, asthma, cardiovascular disease, diabetes, and obesity.[19] Environmental exposures of special interest in these studies include toxins, stress, nutrition, neighborhood, and socioeconomic status (see chap. 10 of this volume). While some of this research directly studies molecular-level mechanisms of gene regulation, most involves epidemiological research on exposures that are hypothesized to interact with the human genome via epigenetic mechanisms but have not yet been experimentally shown to do so in humans.

Epigenetics and Postgenomics

Epigenetics is a postgenomic science. After the completion of the Human Genome Project, epigenetics emerged as a focus for institutional investment and a hot new research area.[20] The early 2000s saw the launch of national and international human epigenome consortia. In 2010, the National Institutes of Health allocated $190 million to the U.S. Human Epigenome Project, promoting it as a "big science" project to succeed the genome sequencing projects.[21]

Epigenetics is shaping public understanding of the promises and prospects of postgenomic science. From pop science books such as *The Epigenetics Revolution: How Modern Biology Is Rewriting Our Understanding of Genetics, Disease and Inheritance* (2012), *Epigenetics: The Ultimate Mystery of Inheritance* (2011), and *The Genie in Your Genes* (2007) to the NOVA television series "The Ghost in Your Genes" (2008) and the *Time* magazine cover article "Why DNA Isn't Your Destiny," epigenetics has met celebratory popular reception.[22] This reception is characterized variously by mysticism and poetic wonder, humor, and hype. Popular writing on epigenetics often projects very nascent scientific findings into practical health advice for everyday life.

Scientists position epigenetics as a bridge between basic genomics research and clinical and public health applications. Pronouncements by scientists about the social, policy, and public health import of epigenetics convey a confidence in the solidity of new epigenetics findings and a conviction that they might be soon applied beyond the laboratory. Ambitious visions of epigenetics as a transformative framework for genomics and public health are nowhere more powerfully developed than in the pronouncements of those working on brain and behavior, such as the McGill University epigeneticists Michael Meaney and Moshe Szyf. As Meaney has written, epigenetics is "likely to have profound consequences when you start to talk about how the structure of society influences cognitive development. We're beginning to draw cause-and-effect arrows between social and economic macrovariables down to the level of the child's brain. That connection is potentially quite powerful."[23] Meaney envisions a science of epigenetics that will show, at the molecular level, the fine-grained biological effects of early stress, deprivation, and trauma and provide support for social policies to reduce these harms. Reflecting this conviction, a conference policy statement from the 2012 San Francisco symposium "The Contribution of Epigenetics in Pediatric Environmental Health" concluded that epigenetics research must be conveyed to the public with the "focused goal" of making public policy changes "faster for the sake of children's health."[24] Szyf extends the implications of epigenetics to the unification of the natural and social sciences and the resolution of the nature-nurture problem. As Szyf writes, "Epigenetics will have a dramatic impact on how we understand history, sociology, and political science. If environment has a role to play in changing your genome, then we've bridged the gap between social processes and biological processes. That will change the way we look at everything."[25]

Scholars in history, philosophy, and social studies of science echo many of these hopes and ambitions. Social scientists analyzing emerging genomic research position epigenetics as a quintessential postgenomic science. They also suggest that it offers a welcome alternative to the genetic determinism of past eras. Joan Fujimura, in her influential 2005 analysis of emerging trends in systems biology, characterized postgenomics as heralding a holist and antireductionist turn in the biosciences. As she wrote, fields such as epigenetics are part of a rising tide of "new 'postgenomic' knowledges that aim to be more ecological and 'wholistic' than the reductionist genetics of the last forty years."[26] Adele Clarke et al. situate epigenetics similarly in the 2010 *Handbook of Genetics and Society*. As they write,

> Rather than genetics revealing a deep, inner, causal truth (a conventional historical assumption), contemporary genetics is instead beginning to conceptualize a 'flattened world' of complex, relayed, dynamic systems of networks of gene-gene interactions, gene-environment interactions, and highly individualized gene expression and regulation that together produce future bodily states. . . . Such . . . conceptualizations also potentially counter some claims about how 'deterministic' genetic and genomic information would detrimentally transform identities. . . . Overall, then, more deterministic outlooks on the impact of genetics are giving way to analyses that emphasize the networked complexities characteristic of the causal models currently used by genetic researchers.[27]

Philosophers of biology, long interested in alternatives to determinist and reductionist explanations in the molecular life sciences, also figure epigenetics as heralding potentially transformative conceptual developments in biology. Philosopher Karola Stotz writes, "A new epigenetic understanding of development encompassing the organism in its developmental niche takes seriously the idea that all traits, even those conceived as 'innate', have to develop out of a single-cell state through the interaction between genetic and other resources of development. Such a view should . . . provide us with a real postgenomic synthesis of development, evolution, and heredity."[28] Biologist-philosopher Eva Jablonka is another among the prominent intellectuals cheering the postgenomic "'move in consensus' in evolutionary biology" toward a "revival of an approach that gives explanatory primacy to development," a move she paints in her book *Transformations of Lamarckism* (2011) as exemplified by the field of epigenetics.[29] In *The Mirage of a Space between Nature and Nurture* (2010), Evelyn Fox Keller similarly suggests that new postgenomic research fields such as epigenetics, systems biology, and studies of phenotypic plasticity portend a postgenomic life sciences trending toward an appreciation of complexity, offering an alternative to the old determinist, reductionist, "particulate" explanatory paradigms of genetics. As Keller concludes, in contrast to "the particulate gene that we inherited from the early days of genetics . . . the new science of genetics coming out of today's research laboratories may point us to a route out."[30]

These accounts render epigenetics as an exemplar of a particularly hopeful conception of postgenomics. Responding to what anthropologist Margaret Lock has called "the lure of the epigenome," science studies scholars pose postgenomics as formed around a disillusionment with the genomic view of the world and its accompanying reductionism (see also chap. 2 of this volume).[31] Holism and anti-determinism are, according to these scholars, a central and unifying frame of the biological sciences after the sequencing of the genome. The new scientific interest in epigenetics is interpreted as demonstrating that biologists are finally realizing—as science studies scholars have long emphasized—that it is not all about the gene. Postgenomics, in this view, marks a move away from an informatic view of genes as a "bag of marbles" toward a material or biochemical conception of genes within the context of the whole genome in relation with its environment. A feature of this worldview is a new respect for the complexity, interdependency, and indeterminacy of gene action. In the science studies literature, epigenetics, along with systems biology and fields such as metagenomics (see chap. 4 of this volume), appears as a prominent and especially salutary example of this new non-gene-centric vision of genomics.

However, a close look at claims in one major area of current epigenetics research, maternal-fetal epigenetic programming, suggests a more complex picture of the relationship between epigenetics and genetic determinism. In

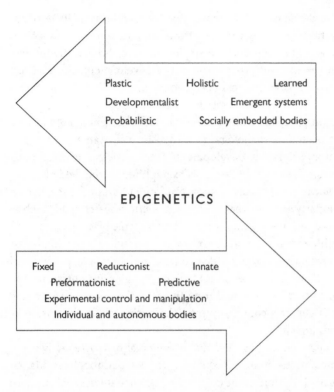

Figure 11.1. Epigenetic explanations combine elements of determinist and anti-determinist frameworks in ways that transcend received conceptual dichotomies and call for new analytic approaches. Author's illustration.

focusing on epigenetics as the alternative to genetic determinism, scholars are operating within a received, twentieth-century dichotomous framework for schematizing forms of biological explanation, one in which a plastic/emergent/socially embedded science of life is positioned opposite a fixed/preformationist/individual and autonomous one (see figure 11.1). This framework, I argue, misreads precisely what is new about the ontological and explanatory terrain of postgenomic sciences such as epigenetics. It also overlooks the significant continuities between the old and the new.

Indeed, determinism and reductionism may not be the most meaningful axes of comparison between "old" and "new" biology. Determinism has to do with the strength and directionality of the causal mechanism; reductionism, with the level of explanation. On examination, while maternal epigenetic effects are not strictly genetic, they are congenital and "fixed" in precisely the same sense as genetically determined traits. They are determinate and predictable in their effects and, potentially, are also heritable. While epigenetic research on maternal effects challenges a notion of the body as the outcome of an indelible genetic code, in its place it offers an expanded but still reductionist and determinate model of development—a "somatic determinism."[32]

Maternal-Fetal Epigenetic Programming

Maternal-fetal epigenetic programming is a thriving area of current human epigenetics research that investigates how exposures during the prenatal and perinatal periods can induce long-lasting epigenetic changes that lead to adult disease and are potentially passed on to future generations. This research field, which is the focus of my discussion here, has come to be known as the "developmental origins of health and disease," or DOHaD. Interest in the effects of undernutrition during gestation on adult disease drove the formation of the field. In the 1980s and 1990s, David Barker, a professor of clinical epidemiology at the University of Southampton, published a series of retrospective birth cohort studies relating prenatal undernutrition and low birth weight to later metabolic and heart disease.[33] Since then, research into the fetal origins of disease has expanded into a large area of study of the "fetal environment." Today, aspects of the fetal environment such as maternal stress and maternal obesity are of especially keen interest.

DOHaD researchers believe that prenatal maternal effects influence development by introducing epigenetic modifications. They hypothesize that the prenatal period is a critical one for "epigenetic programming," in which set points for gene expression are imprinted on the fetus, modifying developmental pathways in areas such as metabolism and the brain. The maternal body, in turn, is conceptualized as an adaptive environment for the fetus in which crucial early developmental cues are transmitted to the growing infant. Because this programming can imprint on a growing female fetus's own gametes as well, the effects of the maternal environment may be intergenerational, passed through the maternal line to grandoffspring.

DOHaD is only one of the major current research areas employing epigenetics frameworks and explanations. The field of epigenetics is in principle quite large in scope. It includes many areas of basic and applied life science research, ranging from core processes of growth and development to the characterization of cancer tumors. These studies may have little connection to maternal-fetal epigenetic programming. Moreover, epigenetics is not the sole focus of all DOHaD researchers. Not all "developmental origins" or "fetal origins" research presumes a strictly epigenetic mechanism for the phenomena under investigation. However, DOHaD, as well as closely allied basic and nonhuman model organism research on maternal epigenetic effects, holds a special status within postgenomic epigenetic science. Theories, data, and experimental paradigms arising from studies of maternal effects have at this time become canonical to the science of epigenetics writ large. There are at least three reasons for this.

First, researchers believe that the prenatal and early developmental stages of life are critical for epigenetic programming. The kinds of stable, long-term epigenetic changes of interest to researchers studying humans are most observable and susceptible to induction and manipulation during the perinatal period. Second, maternal-fetal epigenetic programming intersects with urgent public health priorities in the areas of infant mortality, early childhood development, and prevention of complex and resource-intensive public health problems such as obesity, diabetes, and cardiovascular disease. This presents a rich translational context for attracting investment and interest in epigenetics research. Third, maternal epigenetic effects are central to the most far-reaching and intellectually riveting claims about the philosophical paradigm shift represented by epigenetics. It is in the area of parental epigenetic effects that the prospect of nongenetic intergenerational inheritance becomes a tractable question. It is unsurprising, therefore, that the leading textbook examples of epigenetic effects, such as Meaney's experiments, focus on the role of the maternal body in inducing epigenetic change as the paradigm for postgenomic human epigenetic research.

Research on maternal effects is a new field that over the past decade has grown from a small research stream into a large scientific domain with an international profile. The field now boasts its own journals, specialty workshops, and a scientific society, the International Society for Developmental Origins of Health and Disease, which holds a biennial world congress. DOHaD research has been rapidly incorporated into textbooks and medical curricula and has been particularly influential in public health circles. The field is methodologically and disciplinarily diverse, bridging nutrition science, evolutionary biology, and reproductive physiology, and is host to a variety of perspectives and internal debates. Nonetheless, there exists a core of objectives, empirical claims, language, and conceptual frameworks that distinguish this research field and that may be accessed by surveying its primary literature. Here I profile the theories of two well-known DOHaD researchers.

University College London pediatrician and child nutrition expert Jonathan Wells is an influential theorist of early developmental programming in the prenatal maternal environment. Wells believes that metabolic disorders and obesity are somatic manifestations of the intergenerational transmission of health inequalities. Wells mobilizes DOHaD theory to model how features of the mother's social and environmental context during her own development—including social class—may be transmitted to the growing fetus, conditioning it for a life of inequality even before birth. According to Wells, the maternal body serves as a "transducing medium" for health inequalities from one generation to another.[34] "Maternal capital," as he terms it, is corporealized in the maternal-fetal relation. A diagram of Wells's maternal capital model, reprinted in figure 11.2, schematically illustrates Wells's conception of how public health policies, including education and health care, may be transmitted through the "somatic capital" of the mother to offspring.[35]

While the maternal body can transmit positive resources to the fetus, Wells focuses on "exploring different pathways by which maternal biology may generate ill-health in the offspring."[36] As Wells explains, "offspring are exposed in early life to the 'magnitude of maternal capital.' In any given environment and population, mothers may vary substantially in their capital."[37] Provocatively, Wells has dubbed "low-capital maternal environments," characterized by social and nutritional stressors and other forms of health disadvantage, "metabolic ghettos."[38] As he writes, "While the ghetto in its traditional sense reflects a form of social isolation, I want to extend this concept to a physical bodily dimension and use it to express the impact of economic marginalization on the physiology of reproduction. If pregnancy is a niche occupied by the fetus, then economic marginalization over generations can transform that niche into a physiological ghetto where the phenotypic consequences are long-term and liable to reproduction in future generations."[39] The maternal body in the light of Wells's vivid metaphors of maternal "capital," "ghettos," and "transduction" is a vector that converts maternal social conditions into epigenetic marks on the infant's genome. As an intergenerational vessel of socially inscribed resources that condition life outcomes, the maternal body represents the past, capable of trapping the growing fetus in somatic conditions of deprivation that reproduce social class in postnatal life.

Northwestern University anthropologist Chris Kuzawa is a prominent, highly cited theorist of maternal-fetal epigenetic programming. Trained in evolutionary biology, Kuzawa is interested in the adaptive life history conditions of human metabolism. He researches the prenatal and early-life

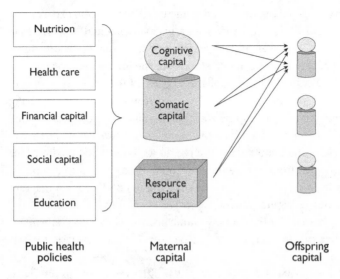

| Public health policies | Maternal capital | Offspring capital |

Figure 11.2. The intergenerational inheritance of "maternal capital." Credit: Wells, "Maternal Capital and the Metabolic Ghetto." Permission to reprint courtesy of John Wiley and Sons. ©2009 Wiley-Liss, Inc.

origins of human metabolic diseases such as diabetes and other obesity-related health conditions. Kuzawa hypothesizes that maternal cues to the fetus, transmitted by hormones and nutrients, can alter adult phenotypic outcomes. Mechanisms such as epigenetics permit plasticity in how an organism realizes its genetic endowment in response to its developmental environment.[40]

These mechanisms are *adaptive* when the mother provides the fetus "access to a cue that is predictive of its future nutritional environment," but they may also be maladaptive.[41] According to Kuzawa, the mother avoids giving maladaptive miscues through a process that Kuzawa labels "intergenerational phenotypic inertia." Epigenetic mechanisms allow the maternal body to transmit a reservoir of signals to the fetus from past maternal ancestors, an inertial adaptation that prevents too great and too rapid phenotypic plasticity. As Kuzawa hypothesizes, "The flow of nutrients reaching the fetus provides an integrated signal of nutrition as experienced by recent matrilineal ancestors, which effectively limits the responsiveness to short-term ecologic fluctuations during any given pregnancy."[42] Continues Kuzawa, "Intergenerational influences on fetal nutrition and growth may act as a form of what may be called intergenerational phenotypic inertia, in the sense that the growth response of the fetus to abrupt ecologic change is tempered by the collective nutritional experiences of recent matrilineal ancestors. Because *the fetal nutritional signal reflects the mother's chronic nutrition tracing back to her own uterine environment, and thereby to prior generations of the matriline*, this may allow the fetus to 'see' an average nutritional environment as sampled over decades and even generations."[43] Kuzawa likens this to a "crystal ball" that allows "the fetus to predict the future by seeing the past, as integrated by the soma of the matriline."[44] Problems arise when this fetal environment for developmental modification and "fine tuning" is either "impaired" or mismatched with current environmental conditions.[45] Maternal epigenetic cues during the prenatal period, he writes, "might be likened to an 'ontogenetic bottleneck' through which any adult metabolic traits must first pass."[46]

Maternal Bodies as Epigenetic Vectors

In DOHaD research, the maternal body emerges as what I call an "epigenetic vector," an intensified space for the introduction of epigenetic perturbations in development. I use the term "vector" here to invoke a series of associations that point to forms of causality that are conduit-like rather than strictly cause-effect, directional rather than distinctly determinative, and relational rather than cleanly linear. In epidemiology, vectors carry disease-causing agents from one organism to another. In aviation, vectoring is a synonym for guiding or directing. In mathematics, vectors determine the position of one point in space in relation to another. Here I identify four critical elements of this vector-like model of explanation and intervention in maternal epigenetic programming research.

First, DOHaD research principally advances a model of epigenetic modifications as a source of error, adverse effects, or disease risk. We might term this a *deficit* model of the relationship of epigenetics to disease. While scientists acknowledge that epigenetics may also provide a route to human enhancement or therapy, at this time the central matter of concern is how to prevent the adverse effects of impaired or maladaptive maternal environments that cause epigenetic lesions in human lineages.

Second, maternal bodies are the central targets of epigenetics-based health intervention. Writes Wells, "Public health policies must be developed to aid the beneficial accumulation of somatic capital and metabolic capacity across generations. . . . I believe this may be achieved by targeting multiple interventions through the transducing medium of maternal capital."[47] As Kuzawa writes, "From an applied perspective, if a trait like fetal growth is designed to minimize the effects of short-term fluctuations by integrating information across generations, public health interventions may be most effective if focused not on the individual but on the matriline."[48] Note that in this vision, the scope of what counts as a "maternal body" is quite large and includes all premenopausal women. As Wells writes, "Whereas [others] have suggested that the reproducing female should be the primary target of interventions to improve health in the next generation, I argue that it is the total period of development of mothers, including experience in their own early life, that is critical to health in the next generation."[49]

Males provide parental care in many species, and male gametes are also subject to environmental exposures that may affect future generations. Although paternal effects are increasingly recognized by scientists and there exist several studies substantiating the existence of intergenerational effects in mammalian paternal lines, currently the focus in DOHaD research is overwhelmingly on the maternal.[50] Typical is the highly cited 2010 review article by Tie-Yuan Zhang and Michael J. Meaney, "Epigenetics and the Environmental Regulation of the Genome and Its Function," which, as a simple metric, uses the terms "maternal" and "mother" 137 times, the terms "parental" and "parent" 11 times, and the terms "paternal" and "father" 3 times.[51]

Scientists defend this imbalance, arguing that "maternal phenotype clearly has substantially greater capacity to shape offspring phenotype through the processes of pregnancy and lactation."[52] But while the exposure of the fetus to the mother is certainly more intimate than to the father in most placental species, this does not fully explain the neglect of research on paternal effects in this field. In her study *Exposing Men: The Science and Politics of Male Reproduction* (2006), the political scientist Cynthia Daniels persuasively documents how, despite evidence that paternal behaviors and life experience such as alcohol use, smoking, and pesticide exposure can impact the health of offspring from conception, scientific research and public health interventions on fetal harm consistently focus on the mother and minimize paternal effects.[53] This pattern endures in DOHaD research. As Daniels demonstrates, this asymmetry originates in long-standing Western cultural and ideological convictions. This includes, on the one hand, a belief in the vulnerability of female bodies and the primary liability of the mother for infant care and development and, on the other hand, a resistance to notions of male reproductive vulnerability and to paternal responsibility for the development of embryos and infants.

Third, while the target of intervention is the maternal body, the desired outcome of epigenetics-driven health interventions is improved fetal

Figure 11.3. *The Fetal Matrix* (2005), by Peter Gluckman and Mark Hanson. Credit: Gluckman and Hanson, *Fetal Matrix*.

health. DOHaD researchers hope that a collateral effect of their policies will be to enhance resources for pregnant women. However, their proposed interventions are directed toward the most efficient methods to ensure developmentally optimum outcomes for the fetus. The symbols favored by DOHaD researchers—on the insignia of its international society, or the cover of one of the field's leading textbooks, *The Fetal Matrix* (2005; figure 11.3)—are fetuses encapsulated in headless, legless maternal abdomens.[54] The maternal body is a transducing and amplifying medium necessary to get to the fetus, an obligatory passage point, not a primary endpoint or subject of DOHaD research.

The possibility of postnatal interventions to reverse maladaptive prenatal epigenetic programming, which would bypass the maternal body, is under investigation. Kuzawa reports, for example, that "the finding that epigenetic changes are durable does not imply that, under changed conditions, the impacts could not be fully or partially reversed. For example, HPA-axis programming in response to maternal care in mice is reversed if offspring are fed a diet supplemented with methyl donors as adults. Similarly, some of the negative metabolic effects of prenatal undernutrition in mice can be reversed by exposure to the fat-derived hormone leptin immediately after birth."[55] While avoiding the maternal conduit, these envisioned interventions to reverse epigenetic programming nonetheless reinforce the message that the field's primary focus is on fetal, not maternal, outcomes. Moreover, dietary or hormonal supplementation to reverse prenatal programming later in life is considered high risk and difficult to test in humans. Researchers express a sense of resignation that the most powerful point of intervention in humans will likely be prenatal. As Wells writes, in humans "post-natal interventions may be incapable of halting the process, since the adaption has already commenced prior to birth."[56]

Fourth and last, while maternal bodies are conceptualized as having great power to influence future generations and are positioned at the center of the intervention model advanced by DOHaD, the DOHaD model accords individual women very little power to influence their own outcomes. On the one hand, women are instructed to do all they can to prevent harm to their fetus. At the same time, an individual woman can do little to improve outcomes for her own offspring if they are trapped in the intergenerational epigenetic "feedforward cycle" hypothesized by DOHaD research. In the DOHaD model, adverse effects originate either in a misalignment between a fetus's inertial epigenetic programming and her eventual environment or in the failure of the maternal environment to provide the necessary epigenetic cues for normal development. In the first scenario, the mother herself is not necessarily the culpable or causal agent. As Kuzawa writes, "If a system is designed to develop with the benefit of ecologic information integrated across multiple generations, short-term treatments in any single generation may reap limited long-term benefits. For such conditions, the most effective focus for intervention may not be the individual but the matriline."[57] The mother's body is a genetically programmed, evolutionarily adaptive system that transmits signals from the past and present environments to the fetus through epigenetic mechanisms at the level of maternal-fetal physiology. This system contains its own inertia. It is beyond conscious or individual control. Change manifests at the level of the intergenerational lineage rather than the individual female. The significance is that DOHaD research advances a shifting and mixed message regarding maternal agency and responsibility: it exhorts mothers to make lifestyle changes in the service of their genetic lineage, while maintaining that these changes are unlikely to bring them or their offspring any benefit. At the same time, it produces a model of the maternal body that suggests that maternal experiences, exposures, and behaviors may have very significant, amplified consequences for her offspring, her descendants, and society at large.

The multivalenced concept of a "vector," perhaps even more precisely than Wells's metaphor of "transduction," points to the distinctive causal-mechanistic explanatory landscape of postgenomic epigenetic science. As an epigenetic vector, the maternal body is at once a background element, a medium for the fetus. Yet it is also a "critical" developmental context in which environmental exposures are amplified, cues are transmitted, and genes are programmed. In epigenetic explanations, elements of agency, control, and intervention mix ambiguously with models of nondirective, inertial developmental systems. Nonetheless, *genes remain very much at the center.* Environments—nutrition, toxins, social policy, stress—are collapsed into molecular mechanisms acting at the level of the DNA. As Kuzawa puts it, epigenetic mechanisms are like "volume controls for genes."[58] It is genes, ultimately, that are expressed and regulated by these epigenetic mechanisms and that create the phenotypic outcomes of concern in this scientific field.

Genetic Determinism and Reductionism in the Postgenomic Order

Epigenetic research on maternal effects advances a model of human inheritance and development in which the wider social and physical environment can be seen as heritable and as a determinate of future biomedical outcomes via discrete biochemical modifications introduced by the amplifying vector of the maternal body. This model is crisply epitomized by a *New Yorker* illustration from a 1997 article on fetal origins research, in which a vector of future developmental outcomes extends outward from the fetus in the woman's abdomen (figure 11.4).[59] Rather than challenging genetic determinism and biological reductionism, it is more precise to observe that present-day research programs in human epigenetics strategically appropriate and modify these discourses to include a particular conception of the social determinants of health, one that places the maternal-infant relation at the center.

In his insightful 2011 article "Epigenetics: Embedded Bodies and the Molecularisation of Biography and Milieu," anthropologist Jorg Niewöhner suggests that epigenetics may make possible the truly biosocial, contextualized science of life that biologists and social scientists have long sought. Niewöhner terms this the "embedded body." Elaborating the current science

Figure 11.4. The maternal body as epigenetic vector.
Credit: Laurent Cilluffo illustration from S. Hall, "Small and
Thin." Permission to reprint courtesy of Laurent Cilluffo.

studies consensus on the paradigm-shifting significance of epigenetics,
Niewöhner presents the embedded body as a novel and exciting alternative to the classic Western biomedical model of the body as mechanistic,
genetically determined, and autonomous. But close attention to Niewöhner's articulation of this "emergent phenomenon . . . made plausible by
environmental epigenetics" reveals certain constitutive exclusions common to many science studies accounts of the theoretical innovations of
epigenetics.[60] As Niewöhner writes, "Epigenetics produces an 'embedded
body', that is, a body that is heavily impregnated by its own past and by the
social and maternal environment within which it dwells. It is a body that
is imprinted by evolutionary and transgenerational time, by 'early-life' and
a body that is highly susceptible to changes in its social and maternal environment. This notion of the body differs significantly from the individual
body with its notion of skin-bound self and autonomy. . . . It suggests an
altogether different degree of entanglement between body and 'context.'"[61]
Niewöhner's account offers a highly discerning analysis of contemporary
epigenetics yet never comments on the prominent figure of the maternal
body in epigenetic theory. The embedded body, according to Niewöhner,

is ever changing, entangled, emergent, and imprinted by its environment. Indeed, it appears that the "embedded body" is an archetypal fetus. In turn, the environment is embodied by the fuzzy, receding figure of the maternal.

An appreciation of the figuring of the maternal body as a vector in epigenetics leads to a different reading than Niewöhner's. Epigenetics does not so much "make plausible" the embedded body; rather, it fixes the molecular gaze on the embedded body, an already-formed and highly charged entity in the science of maternal-fetal relations, and elevates it to the center of biomedical theory, intervention, and surveillance.[62] Here epigenetics does not so much "entangle . . . bodies and contexts"; rather, it brings the "environment"—transduced through the maternal body—into processes of biomedicalization, optimization, and manipulation of life initiated by the twentieth-century molecular life sciences.[63]

Scientists and science studies scholars position epigenetics as a holist, antireductionist, anti-determinist counterdiscourse to the genetic determinism of twentieth-century genetics. Yet models of causality, mechanism, and intervention at work in maternal-fetal epigenetic programming research suggest that this dichotomous opposition between epigenetics and genetic determinism is inadequate to appreciate the conceptual shifts and social, ethical, and philosophical challenges introduced by postgenomic ventures such as epigenetics. Reflection on epigenetics-based biomedical and public health interventions recommended by scientists such as Wells and Kuzawa suggests a need for science analysts to shift the critical terms of debate away from a discussion driven by a concern over the harms of genetic determinism to one more sensitive to how certain bodies or spaces become intensive targets of intervention when conceptualized as amplifying vectors of developmental or epidemiological risk.

The ground has shifted under science analysts' feet. The terms of debate— the relevant distinctions and differences that ordered theoretical approaches to molecular genetics—have been reordered. What is interesting and critical is not the assessment of epigenetics in terms of old oppositional fault lines of determinism versus redundancy and pleiotropy, reductionism versus holism and complexity. This old framework was built to analyze a different genomic moment. In collapsing the oppositions of the old, the postgenomic order, to the extent that it is continuous with the explanatory landscape of epigenetics, breaks received binaries and calls for new conceptual maps.

Gender analysis of knowledge production in maternal-fetal epigenetic programming research suggests that the postgenomic order, as I call it, is one that blurs, reforms, or breaks down the received oppositional binaries of the molecular genetics era depicted in figure 11.1. This is an exciting agenda for science studies scholars in the postgenomic era. To access the new ontology, we might heed Karen Barad's calls for a "diffractive" rather than merely "reflective" reading of the "entanglements" that produce knowledge claims. Diffractive reading, in Barad's formulation, focuses on "what differences matter" and where the "agential cuts" are made within an explanatory apparatus.[64] The case of the maternal body in epigenetics hints that hybrid assemblages comprising vectors, spaces, boundaries, amplifications, temporalities, and hot spots may point to the foundations of the explanatory and analytical language of the postgenomic order.

Conclusion

The case of the maternal body in epigenetics research reveals gender as an analytical frame for examining epigenetics with fresh eyes and from a new vantage point. The foregrounding of the maternal body as an epigenetic vector in postgenomic biomedical research resonates with the history of highly politicized conceptions of maternal responsibility and may further extend biomedical manipulation and social control of the reproductive female body. Gender analysis of science, however, offers insights that extend beyond investigation of the consequences of scientific research for women. By attending to the production of embodied differences, gender analysis can help to reveal theoretical shifts instantiated in the postgenomic age not apparent through standard analytical frames. Abandoning simple views of the gene and of gene action, and providing a mechanism for gene-environment interactions, the burgeoning field of epigenetics attests to what some see as the hallmark of postgenomics—a disillusionment with a reductionist molecular genetic view of the world. Yet, as I argue in this chapter, in its focus on biochemical modifications of DNA mediated by the maternal body, epigenetic science reproduces and subtly reformulates determinist strategies and reductive methods. The figure of the *maternal body as epigenetic vector*, I wish to suggest, can be seen as one highly enriched ensign of the explanatory landscape of an emergent postgenomic order that yet lies beyond the grasp of our preformed categories.

Notes

This research was supported by fellowships from the American Council of Learned Societies and the Radcliffe Institute for Advanced Study and a seed grant from the Robert Wood Johnson Foundation. Alex Margarite and Sean Nunley provided research assistance.

1. Meaney, "Maternal Care, Gene Expression."

2. Blanchard, "Fraternal Birth Order."

3. Paul, *Origins*.

4. See, e.g., Ladd-Taylor and Umansky, *"Bad" Mothers*; Plant, *Mom*; Warner, *Perfect Madness*.

5. Parke, *Century of Developmental Psychology*.

6. See, e.g., Casper, *Making of the Unborn Patient*; Rapp, *Testing Women, Testing the Fetus*.

7. For exemplary feminist discussions of these issues in the case of maternal-fetal epigenetic programming research, see, e.g., McNaughton, "From the Womb to the Tomb"; Warin et al., "Mothers as Smoking Guns."

8. Ladd-Taylor and Umansky, *"Bad" Mothers*; Kukla, *Mass Hysteria*.

9. Schramek, "Michael Meaney"; Sebald, *Momism*.

10. Cynthia R. Daniels, "Between Fathers and Fetuses."

11. Lock, "Eclipse of the Gene"; Griesemer, "Turning Back to Go Forward"; Haig, "Weismann Rules!"

12. McGowan et al., "Epigenetic Regulation of the Glucocorticoid Receptor."

13. For an excellent review of epigenetic mechanisms see Mehler, "Epigenetic Principles and Mechanisms."

14. Morgan et al., "Epigenetic Inheritance at the Agouti Locus in the Mouse."

15. Lee and Zucker, "Vole Infant Development."

16. Szyf et al., "Maternal Programming of Steroid Receptor Expression."

17. During meiosis (the division process that generates the sperm and egg cells) and early embryo development, chromosomes are restructured and embryonic DNA experiences a genome-wide demethylation. On this basis, it is generally believed that environmentally induced epigenetic changes are not inherited through the germ line and cannot be directly passed from parent to offspring.

18. The most widely discussed body of data regarding the long-term intergenerational effects of early developmental exposures comes from the so-called Dutch Hunger Winter studies. A useful review of these studies is given by Roseboom, "Undernutrition during Fetal Life." For a sense of the contestations over how to interpret these findings, see, e.g., S. Hall, "Small and Thin."

19. Rodriguez et al., "Maternal Adiposity Prior to Pregnancy"; Badcock and Crespi, "Imbalanced Genomic Imprinting"; Bocklandt et al., "Extreme Skewing of X Chromosome Inactivation"; Wright, "Prenatal Maternal Stress and Early Caregiving Experiences"; Dyer and Rosenfeld, "Metabolic Imprinting by Prenatal, Perinatal, and Postnatal Overnutrition"; Barker, "Fetal Origins of Coronary Heart Disease."

20. Feinberg, "Epigenetics at the Epicenter of Modern Medicine"; Petronis, "Epigenetics as a Unifying Principle."

21. Pennisi, "Are Epigeneticists Ready for Big Science?"

22. Carey, *Epigenetics Revolution*; Francis, *Epigenetics*; Church, *Genie in Your Genes*; Holt and Paterson, "Ghost in Your Genes"; Cloud, "Why Your DNA Isn't Your Destiny."

23. Weaver, Meaney, and Szyf, "Maternal Care Effects."

24. Children's Environmental Health Network, "Contribution of Epigenetics in Pediatric Environmental Health."

25. Weaver et al., "Maternal Care Effects."

26. Fujimura, "Postgenomic Futures."

27. Clarke et al., "Biomedicalizing Genetic Health, Diseases and Identities," 26–27.

28. Stotz, "Ingredients for a Postgenomic Synthesis of Nature and Nurture," 362.

29. Jablonka, "Lamarckian Problematics in Biology," 145.

30. Keller, *Mirage of a Space.*

31. Lock, "Lure of the Epigenome."

32. Ibid., 1896.

33. Wadhwa et al., "Developmental Origins of Health and Disease." See also Barker, "Fetal Origins of Coronary Heart Disease"; Gluckman and Hanson, *Fetal Matrix*, and *Developmental Origins of Health and Disease.*

34. Wells, "Maternal Capital and the Metabolic Ghetto," 13–14.

35. Ibid.

36. Wells, "Thrifty Phenotype," 145.

37. Wells, "Critical Appraisal of the Predictive Adaptive Response Hypothesis," 230.

38. Wells, "Maternal Capital and the Metabolic Ghetto"; Wells, "Thrifty Phenotype," 163.

39. Wells, "Maternal Capital and the Metabolic Ghetto," 11.

40. Kuzawa, "Why Evolution Needs Development," 224.

41. Kuzawa, "Fetal Origins of Developmental Plasticity," 5.

42. Ibid.

43. Ibid., 12; emphasis added.

44. Ibid., 10, 13.

45. Kuzawa, "Why Evolution Needs Development," 226; Kuzawa, Gluckman, and Hanson, "Developmental Perspectives on the Origins of Obesity," 207.

46. Kuzawa et al., "Developmental Perspectives on the Origins of Obesity," 211.

47. Wells, "Maternal Capital and the Metabolic Ghetto," 13–14.

48. Kuzawa, "Fetal Origins of Developmental Plasticity," 5.

49. Wells, "Thrifty Phenotype," 165.

50. Ibid., 154; Pembrey et al., "Sex-Specific, Male-Line Transgenerational Responses in Humans"; Shulevitz, "Why Fathers Really Matter."

51. Zhang and Meaney, "Epigenetics and the Environmental Regulation."

52. Wells, "Thrifty Phenotype," 154.

53. Daniels, *Exposing Men.* On the neglect of paternal effects, see also Armstrong, *Conceiving Risk, Bearing Responsibility.*

54. Gluckman and Hanson, *Fetal Matrix.*

55. Thayer and Kuzawa, "Biological Memories of Past Environments," 801.

56. Wells, "Thrifty Phenotype," 161.

57. Kuzawa, "Fetal Origins of Developmental Plasticity," 17.

58. Kuzawa and Sweet, "Epigenetics and the Embodiment of Race," 5.

59. S. Hall, "Small and Thin."
60. Niewöhner, "Epigenetics," 290.
61. Ibid.
62. See also Landecker, "Food as Exposure."
63. See Clarke et al., "Biomedicalizing Genetic Health, Diseases and Identities."
64. Barad, *Meeting the Universe Halfway.*

12

Approaching Postgenomics

Sarah S. Richardson and Hallam Stevens

Will future scholars divide biology into "pre- and post-genomic eras"? Is a grand conceptual rupture under way, a progressive transition to a new, more advanced age of "Biology 2.0"?[1] Any such assertions are challenged by the close scrutiny of recent biology offered by the contributors to this book. It is clear that there are both continuities and discontinuities between the eras of molecular biology, genetics, genomics, and postgenomics. The tension between the new and the old lies at the heart of what postgenomics is and how it will develop.[2]

The first uses of the term "postgenomics" can be traced to a series of anticipatory meetings held by Human Genome Project (HGP) scientists in the 1990s known as the "After the Genome" conferences.[3] The consensus statements produced by these convenings reveal remarkably prescient anticipation of the shifts that would be brought about by the availability of whole-genome data and technologies. The conference statement produced by attendees of the 1999 "After the Genome IV: Envisioning Biology in the Year 2010" meeting predicted that postgenomics would comprise the study of interacting systems such as proteomics, studies of gene-environment interactions, and increasingly complex models of biological pathways requiring powerful and sophisticated informatics and computational skills, as well as expanded and speedier sequencing of genomes. As the authors wrote, "Perhaps by 2010, biology will, in fact, have reached the end of the reductionist road, and efforts will be largely directed at reassembling the

pieces that took more than half a century to dissect into a more compre-
hensible whole. In such an environment, meetings like this will become
more frequent, especially if the various computer simulations promising to
predictively integrate genetic sequence, expression, and physiological data
begin to make good on their promises."[4]

Today, the goals of postgenomic biology include the study of the relationship between genotype and phenotype (the broadest sense of the term
"epigenetics"); proteomics, or the characterization of the proteins encoded by DNA sequence, including their structure, expression profile, and
interactions; and systems biology, integrating sequence, proteomics, and
structural and functional genomics into models of normal and pathological
organismic biology. Postgenomics appears as a set of approaches, dubbed
the "-omics," that extend existing genomic programs and paradigms across
the many subfields of the life sciences. In this vision of postgenomics,
genomics operates as a unifying framework for biological knowledge, and
postgenomics represents—in Kuhnian terms—the puzzle-solving mode of
the genomics era. Notably, the genome itself remains the central object:
common metaphors describe genomics as a "foundation" for the study of
organismic biology and evolution, a "scaffold" for understanding biological
systems, and a "catalogue" or "database" of functional biology.[5]

Indeed, the enthusiasm for genes and genomes appears to be as great as
ever. In 2010, the 1000 Genomes Project published the first results of its
initiative to use large-scale sequencing to identify all variants in the human
genome that have a frequency greater than 1 percent.[6] In 2012, the U.K. National Health Service announced plans to sequence 100,000 British genomes
and match them against patients' health records.[7] Also that year, the consumer genome sequencing company 23andMe raised $50 million in venture
capital and dropped its price for a full individual genome report to just $99,
projecting one million users in the coming decade.[8] Today, automatic genome
sequencers are cheaper and faster than ever, with the most recent versions
of Illumina machines capable of producing one hundred gigabases (or thirty
human genomes' worth) of data per day.[9] Researchers aim to sequence tens
of thousands of whole genomes of humans and nonhumans. BGI (formerly
Beijing Genomics Institute) in Shenzhen, China, claims to have already sequenced 50,000 human genomes.[10] In addition to this vast amount of human
DNA, by 2013 almost four thousand bacterial and archaea species had been
completely sequenced, as well as almost two hundred eukaryotes, including
organisms as diverse as platypus, elephant, puffer fish, and mosquito.

Likewise, popular interest in the power of genes remains strong. In
2013, *New York Times* headlines announced that a "Mouse Study Discovers

DNA That Controls Behavior" and, in the aftermath of the Newtown school shooting, that the University of Connecticut would study the DNA of the gunman in "an effort to discover biological clues to extreme violence."[11] In recent years, leading American intellectuals and public figures such as Steven Pinker, Henry Louis Gates, and Oprah Winfrey have, by personal example and high-octane media exposure, promoted genome sequencing as a route to self-knowledge and maximum health.[12]

However, postgenomics also shows some clear breaks with earlier modes of work. Increasingly, genome scientists rely on multidimensional computational and informatics approaches demanded by more and new kinds of data. This raises distinctive issues for knowledge validation. As genomicists abandon simple views of gene action, new systems-based frameworks and forms of genomic explanation challenge traditional forms of research. Greater demand by funders and publics for research that quickly translates to the clinic and to the marketplace is also transforming the postgenomic life sciences. Government funders of the molecular life sciences have tightened the reins on funding after the economic crisis and end of the formal funding commitments of the HGP. Now, "translational" science is the watchword for grantmakers such as the National Institutes of Health (NIH), which increasingly demands that genome research show direct links to therapies, clinical applications, or economic development. This shift is reflected in changes in the language of grant applications and throughout the mission statements of the NIH and other agencies.[13] All of these hallmarks of the postgenomic life sciences structure the choices and constraints of genome scientists. These features also have the potential to reenergize and raise new social, political, and ethical questions about the implications of genomic technologies for human communities.

Alongside the exuberant messages of confidence in the genomic methods described above, others express more circumspection. Once-marginal critiques of a gene-centric vision of the life sciences are now moving to the center of the action. Scientists across many fields have suggested that the key to human health and behavior will not be found in genomic code but in the delicate networks of gene regulation in their biochemical, cellular, and ecological environments.[14] The relationship between genes and human traits, genomicists increasingly concede, is significantly more complex than initially envisioned.

These multiple meanings, agendas, and stakes suggest that "postgenomics" cannot be adequately conceptualized as a simple "break" or "transition" from one era to the next. Nor can it be characterized as a straightforward continuation of genetics or genomics. Rather, the essays in this volume have collectively begun to theorize postgenomics in terms of an ongoing struggle to find new ways of thinking, working, and explaining within the parameters set by the tools of genomics. Postgenomics emerges from a tension between the constraints of specific tools and methods and biologists' attempts to draw new sorts of models and findings from them. The rest of this conclusion examines how these tensions are playing out in the domains of scientific explanation (the tension between reductionism and complexity), transdisciplinanty (the tension between experimental and competing forms of knowledge production), and translation (the tension between basic biology and medical applications).

Explanation

The emergence of subdisciplines such as systems biology, as well as increased attention given to epigenetics and gene-environment interactions, seems to signal postgenomic moves toward more holist approaches and away from genetic reductionism. Postgenomic explanations move beyond the explanatory frame of the "gene for X" to include both molecular and environmental signals and to feature patterns and associations, multiple levels of causation, and mechanisms dispersed across networks.

Fields such as microbiomics and metagenomics confute the notion of the human genome as a contained and complete record of the genomic content of the human body. Not only does this research trouble the foundations of the traditional focus on the human genome as the blueprint for human life and disease, but it suggests a further realm of data-rich material for the discovery of differences between individuals and populations. In all of these ways, notions of difference and similarity, individuals and populations, are being actively refigured in the postgenomic age with potentially explosive implications for biological theory and for conceptions of human genomic variation.

Yet genes and sequences still play a central role in postgenomic thinking, and reductionism has not disappeared. "Genome talk" is something different, but not entirely different, from its predecessor "gene talk."[15] Examination of what precisely is being reduced to what, how "holist" approaches operate in particular contexts, and what such models include and exclude suggests that rather than overcoming reductionism, postgenomic explanations meld forms of reductionism and holism in new ways. The forms of reductionism that have emerged in postgenomics present different kinds of elisions and simplifications than in previous eras of molecular biology. Below the surface of postgenomic affirmations of complexity, many research programs remain centered on determinist causal explanations located in small molecular mechanisms at or on the genome.

The contested status of postgenomics knowledge became visible in 2013 in the trials of the personal genome testing company 23andMe. The business provides risk assessments for their customers based on comparison of an individual's personal genotype with results from published genome-wide association studies. The U.S. Food and Drug Administration complained that the company was offering diagnostic tests without regulatory oversight and contended that it was giving people unverified information about genetic correlations.[16] 23andMe's supporters responded by arguing that the company's tests should be no more subject to regulation than the production of thermometers (which might also be used by consumers to make inferences about their own health).[17] As a mere provider of personal health information, they argue, 23andMe should have a special status and should lie outside of the conventional regulatory apparatus for medical testing and diagnosis. This debate raises thorny questions about the explanatory status of genomic correlations. A persistent question for the postgenomic era—one with ramifications ranging from the legal to the philosophical—will be the causal status of such correlations within an increasingly complexity-affirming explanatory repertoire.

More broadly, these tensions can be perceived in the debates about "hypothesis-free" or "data-driven" biology. Here biologists have begun to argue explicitly about what counts as authenticated knowledge. Does a genome-wide association study linking cancer to a particular location on the genome really constitute biological knowledge? Or must biologists do further work in the laboratory to elaborate the mechanism of action before such a finding can count as knowledge? Massive amounts of data, analyzed with new techniques, promise new insights into biology and disease. But many observers (both inside and outside of biology) remain suspicious of results based on correlations, patterns, and associations without clear links to established biological mechanisms.[18] Science studies scholars have an important role to play in understanding the explanatory dilemmas of postgenomic science and developing frameworks for assessing and understanding these new strategies for building knowledge.

Transdisciplinarity

Debates over appropriate forms of explanation are closely linked to disciplinary and institutional shifts within the life sciences. Biological work in the postgenomic age *looks* remarkably different from its predecessors.

Doing biology today requires new kinds of skills, new kinds of work, and new ways of making knowledge. Individuals "doing" biology—in a broad sense—may now have training in fields as diverse as physics, computer science, statistics, informatics, engineering, operations, logistics, and management. The spaces in which biology is done, too, may look much more like offices or factories than traditional labs. Within this dynamic and transdisciplinary landscape, research communities in biology have had to adopt and adjust to new methodologies, to work out what constitutes a valid study design, to decide on what counts as a significant or publishable finding, and to find new ways of validating and authorizing knowledge claims. The emergence of new kinds of work is shifting the ground for receiving credit and recognition in biology.[19]

Computational and bioinformatic forms of biology, in particular, pose a challenge to older ways of knowing and doing in biology. Although some biologists still insist that any finding must be "validated" in a wet-lab experiment, it is increasingly possible to do biology from one's desktop computer. Biologists have had to find ways of working with and alongside statisticians and computer scientists and coming to trust the computers and algorithms that they use. Even within the biological laboratory itself, there has been an increasingly pronounced difference between older forms of experimental work and "high-throughput" experiments that place a premium on rapidity, consistency, efficiency, accuracy, and low cost (rather than traditional forms of scientific creativity or innovation). Indeed, these "production" activities—DNA and RNA sequencing—are at the heart of postgenomics.[20]

Debates about data and hypothesis-free biology are not just about *what* counts as knowledge but also about *who* counts as an authorized knowledge producer: does someone mining a biological database for new associations count as a biologist? Such problems and questions are intricately linked to the technologies and tools of postgenomics. Genomic biology deployed a powerful set of tools—DNA sequencing, databases, and algorithms—for studying genomes. In the postgenomic era, these tools have been developed into next-generation sequencing machines, as well as bigger, faster, and more powerful databases and algorithms. But it is precisely the development of these tools that has necessitated the incorporation of new skills and modes of working into biology: next-generation sequencing required engineers and roboticists, databasing required informaticists, algorithms required computer scientists and mathematicians, and so on. Scaling up the tools of genomics to move beyond older frameworks has necessitated a transdisciplinary orientation that is refiguring the social arrangements underpinning the authentication of biological knowledge claims.

Translation

Part of the new transdisciplinarity of postgenomics is its increasingly bio-medical orientation—that is, much postgenomic biology is directed toward therapeutic and clinical applications. Highly complex environmentally and socially mediated human diseases rank among the greatest economic challenges in Western societies. Today, efforts such as the NIH Gene, Environment, and Health Initiative are major research priorities at the level of Big Science and are receiving substantial financial resources. Ambitious research programs in genomic epidemiology, behavioral genetics, and epigenetics seek to add layers of information about the "environment" to genomic research models.[21]

Within this translational mode, race, gender, nationality, and ethnicity are prominent categories for analyzing human population variation and searching for genotypes relevant to health, human evolutionary history, and personal ancestry. Initiatives such as the International HapMap Project and 1000 Genomes Project aim to sequence whole genomes or haplotypes of racially and ethnically defined populations, expanding sequence data far beyond the initial human reference genome. There is a translational imperative to include diverse races, ethnicities, and nationalities in genome research, informed by the aims of an inclusive and diverse science, the commercial aims of ethnomedicine, and optimistic postracial politics.[22] Yet these imperatives jockey with the continuing need to guard against racial stereotyping in medicine and racist or racialist assumptions in research design and interpretation. Similarly, the ideal of arriving at unique and personalized whole-genome analysis of individuals ambivalently coexists with the continued and necessary practice of population stratification for the derivation of meaningful haplotypes. Refigured in relation to "translational" social justice aims and the amelioration of health disparities, and situated in the context of the globalization of genomic institutions and knowledge production, postgenomic conceptions of human difference will require newly sensitive approaches to the contextual conditions of race-based genomics research.[23]

As the essays in this book demonstrate, rather than merely an external, unidirectional influence on the sciences, the pressure for "translation" is a bidirectional, looping phenomenon: as the results of genomic research are rushed into commercial, clinical, and public health applications, social imperatives and formations are also influencing practices in genomics. The demand for "socially relevant" biology not only generates new modes of working and doing but also requires biologists to develop new modes of engaging with fraught issues such as race and gender. As the translational imperative becomes more powerful in the social and cognitive practices of the postgenomic life sciences, the tools of science studies stand as resources to help understand and assess the complex links between the laboratory and diverse translational contexts, including the clinic.

Beyond the Century of the Gene

There exists a strong body of science studies scholarship analyzing the consequences of genomics for the political, social, ethical, economic, and clinical spheres.[24] This volume, however, has focused its attention on transformations in concepts, practices, and research agendas *within* the life sciences in the era since the arrival of the genome sciences. We hope that this approach may serve to further connect scholarly debates within science studies to the discussions and debates that are taking place within the life sciences themselves. By remaining close to the day-to-day work of biologists, we have aimed to pose questions that life scientists might recognize and engage with in their everyday practices. By highlighting science studies analysis inside the labs, in the primary scientific literature, and within the databases and software of contemporary biology, we hope that these essays show how such work might be of direct relevance to thinking about disciplinary and epistemic problems within biology. Dialogue of this sort is essential, not only to continue pushing biologists to be reflective about their own work but also to challenge social scientists to examine their own assumptions and to sharpen their critical tools as we all navigate the postgenomic era.

In the 1990s, historians, sociologists, anthropologists, and philosophers cautioned against the hyperbolic promises of the genomic era and criticized the reductionism and genetic determinism that seemed to lie at the heart of the HGP. As the sequencing projects were completed, the genome's complexity became increasingly apparent. In the postgenomic period, biologists, perhaps now with greater humility, have paid more attention to holism and to environmental and developmental factors. Social scientists struggling to find a mode of critical engagement with postgenomic biology must appreciate the transformative dimensions of postgenomics— especially its attempts to move beyond the simple models and dichotomies of the "century of the gene."[25] On the other hand, many aporiae remain: postgenomics continues to promise big, genes and sequences still play a central role in postgenomic thinking, and reductionism has not disappeared. The diverse tools of science studies, from laboratory ethnography to discourse analysis, field studies, methodological critique, historical contextualization, and political intervention, offer resources for bringing transparency and empirical precision to descriptions of the practices of postgenomics. Yet, as the case of reductionism and determinism in postgenomics suggests, science studies scholars must be conscious that as the ontologies and social arrangements of the postgenomic life sciences shift and reform, the analytic and critical frameworks of the past may require rethinking and revision.

The genomic sciences are growing in importance as a platform both within and beyond the life sciences. The perceived success of genomics is leading other fields—including domains as diverse as neuroscience, ecology, music, economics, political science, and sociology—to adopt aspects of the genomic approach.[26] This includes the application of genomics methods and elements of genomics practice such as the use of large databases, multidisciplinary teams, data sharing, high-throughput data collection and experimentation, hypothesis-free "discovery," and massive computation. Close attention to the workings of genomics may shed light on why this model of scientific work has become so compelling. But it may also serve as an important indicator of the future: understanding the dynamics of how genomics has transformed over the past decade is going to be crucial in understanding not just how biology will develop but also where the sciences (and perhaps also the social sciences) are headed in the coming decades.

Notes

1. "Biology 2.0."
2. Indeed, reflection on "postgenomics" provides an opportunity to examine how we develop concepts of periodicity in science without recourse to models of abrupt rupture and to critically analyze what purposes are served by the construction of different eras, periods, shifts, or movements in science. See Koselleck, *Practice of Conceptual History*.
3. Brent, "After the Genome IV."
4. Ibid.
5. Lander, "Initial Impact of the Sequencing of the Human Genome." On metaphors of genomics, see Nerlich and Hellsten, "Genomics."
6. Nielsen, "Genomics."
7. Akst, "100,000 British Genomes."
8. Tsotsis, "Another $50M Richer."
9. According to Illumina, their *HiSeq* 2500 produces 120 GB in approximately twenty-seven hours; see http://www.illumina.com/systems/hiseq_comparison.ilmn.
10. Larson, "Inside China's Genome Factory."

11. Gorman, "Mouse Study Discovers DNA"; Kolata, "Scientists to Seek Clues."

12. Maggio et al., *Faces of America*; Pinker, "My Genome, My Self."

13. See National Institutes of Health, "NIH Roadmap"; S. H. Woolf, "Meaning of Translational Research."

14. See, e.g., chap. 2 of this volume.

15. Nelkin and Lindee, *DNA Mystique*, 1st ed.

16. Gutierrez, "Letter to Anne Wojcicki."

17. Wojcicki, "23andMe Provides an Update."

18. For instance, Jon McClellan and Mary-Claire King argued in *Cell* that "to date, genome-wide association studies (GWAS) have published hundreds of common variants whose allele frequencies are statistically correlated with various illnesses and traits. However, the vast majority of such variants have no established biological relevance to disease or clinical utility for prognosis or treatment"; McClellan and King, "Genetic Heterogeneity in Human Disease." See also Turkheimer, "Genome Wide Association Studies."

19. See, e.g., chap. 7 of this volume.

20. Stevens, "On the Means of Bio-production."

21. Shostak and Freese, "Gene-Environment Interaction and Medical Sociology."

22. See, e.g., chap. 9 of this volume.

23. Reardon, "Democratic, Anti-racist Genome?"

24. See, e.g., Bliss, *Race Decoded*; Shostak, *Exposed Science*; Barnes and Dupré, *Genomes and What to Make of Them*; Rose and Rose, *Genes, Cells and Brains*; N. Rose, *Politics of Life Itself*; Sunder Rajan, *Biocapital*; Atkinson, Glasner, and Greenslade, *New Genetics, New Identities*; Fortun, *Promising Genomics*; Koenig, Soo-Jin Lee, and Richardson, *Revisiting Race in a Genomic Age*; Hilgartner, *Science and Democracy*.

25. On this point, see Meloni, "Biology without Biologism"; Niewöhner, "Epigenetics."

26. On economics and political science, see Callaway, "Economics and Genetics Meet"; Hatemi and McDermott, "Genetics of Politics." On music, see the description of the Music Genome Project at http://www.pandora.com/about/mgp. The Human Connectome Project, funded by the NIH, is partly modeled on genomics: http://www.humanconnectomeproject.org/. In ecology, projects such as the MoveBank database (https://www.movebank.org/) are modeled on large biological databases such as GenBank. The emergence of chemoinformatics suggests the impact of genomics on chemistry. See Brown, "Editorial Opinion."

BIBLIOGRAPHY

Abbate, Janet. *Inventing the Internet*. Cambridge, MA: MIT Press, 1999.

Akst, Jef. "100,000 British Genomes." *Scientist*. December 10, 2012. http://www.the-scientist.com/?articles.view/articleNo/33622/title/100-000-British-Genomes/.

Albert, Réka, and Albert-László Barabási. "Statistical Mechanics of Complex Networks." *Reviews of Modern Physics* 74 (2002): 47–97.

Albert, Réka, Hawoong Jeong, and Albert-László Barabási. "The Diameter of the WWW." *Nature* 401 (1999): 130–31.

Albert, Réka, Hawoong Jeong, and Albert-László Barabási. "Error and Attack Tolerance of Complex Networks." *Nature* 406 (2000): 379–81.

Allen, J. F. "*In silico veritas*: Data-Mining and Automated Discovery: the Truth Is in There." *EMBO Reports* 2 (2001): 542–44.

Alon, Uri. "Network Motifs: Theory and Experiment." *Nature Reviews Genetics* 8 (2007): 450–61.

Alper, Joseph S., and Jonathan Beckwith. "Genetic Fatalism and Social Policy: The Implications of Behavior Genetics Research." *Yale Journal of Biology and Medicine* 66 (1994): 511–24.

Altman, Rebecca Gasior, Rachel Morello-Frosch, Julia Green Brody, Ruthann Rudel, Phil Brown, and Mara Averick. "Pollution Comes Home and Gets Personal: Women's Experience of Household Chemical Exposure." *Journal of Health and Social Behavior* 49 (2008): 417–35.

Amin, M. "Modeling and Control of Complex Interactive Networks." *IEEE Control Systems Magazine* 22 (2002): 22–27.

Ankeny, Rachel A., and Sabina Leonelli. "What's So Special about Model Organisms?" *Studies in the History and Philosophy of Science* 41 (2011): 313–23.

Anthony, Sebastian. "Google Compute Engine: For $2 Million/Day, Your Company Can Run the Third Fastest Supercomputer in the World." *Extreme Tech*, June 28, 2012. http://www.extremetech.com/extreme/131962-google-compute-engine -for-2-millionday-your-company-can-run-the-third-fastest-supercomputer -in-the-world.

Armstrong, Elizabeth M. *Conceiving Risk, Bearing Responsibility: Fetal Alcohol Syndrome and the Diagnosis of Moral Disorder*. Baltimore: Johns Hopkins University Press, 2003.

Arnaud, Celia Henry. "Exposing the Exposome." *Chemical and Engineering News* 88 (2010): 42–44.

Arribas-Ayllon, Michael, Andrew Bartlett, and Katie Featherstone. "Complexity and Accountability: The Witches' Brew of Psychiatric Genetics." *Social Studies of Science* 40 (2010): 499–524.

Atkinson, Paul, Peter E. Glasner, and Helen Greenslade. *New Genetics, New Identities*. New York: Routledge, 2006.

Auchincloss, Amy H., Ana V. Diez Roux, Mahasin S. Mujahid, Mingwu Shen, Alain G. Bertoni, and Mercedes R. Carnethon. "Neighborhood Resources for Physical Activity and Healthy Foods and Incidence of Type 2 Diabetes Mellitus: The Multi-Ethnic Study of Atherosclerosis." *Archives of Internal Medicine* 169 (2009): 1698.

Augustin, T., T. A. Glass, B. D. James, and B. S. Schwartz. "Neighborhood Psychosocial Hazards and Cardiovascular Disease: The Baltimore Memory Study." *American Journal of Public Health* 98 (2008): 1664–70.

Badcock, C., and B. Crespi. "Imbalanced Genomic Imprinting in Brain Development: An Evolutionary Basis for the Aetiology of Autism." *Journal of Evolutionary Biology* 19, no. 4 (2006): 1007–32.

Balshaw, David M. "Making the Case for Advancing the Exposome (or EWAS)." National Institute of Environmental Health Sciences, 2010.

Barabási, Albert-László. "Growth and Roughening of Nonequilibrium Interfaces." PhD diss., Boston University, 1994.

Barabási, Albert-László. *Linked: The New Science of Networks*. Cambridge, MA: Perseus, 2003.

Barabási, Albert-László, and Réka Albert. "Emergence of Scaling in Random Networks." *Science* 286 (1999): 509–12.

Barad, Karen Michelle. *Meeting the Universe Halfway: Quantum Physics and the Entanglement of Matter and Meaning*. Durham, NC: Duke University Press, 2007.

Barker, D. J. P. "Fetal Origins of Coronary Heart Disease." *British Medical Journal* 311, no. 6998 (1995): 171–74.

Barnes, Barry, and John Dupré. *Genomes and What to Make of Them*. Chicago: University of Chicago Press, 2008.

Bastow, Ruth, and Sabina Leonelli. "Sustainable Digital Infrastructure." *EMBO Reports* 11 (2010): 730–34.

Bearman, Peter. "Genes Can Point to Environments That Matter to Advance Public Health." *American Journal of Public Health*, e-View Ahead of Print, 2013.

Bearman, Peter, Molly A. Martin, and Sara Shostak. *Exploring Genetics and Social Structure*. Chicago: University of Chicago Press, 2008.

Beckwith, Jon. "Foreword: The Human Genome Initiative: Genetics' Lightning Rod." *American Journal of Law and Medicine* 17 (1991): 1–13.

Bellman, Richard. *Dynamic Programming*. Princeton, NJ: Princeton University Press, 1957.

Belvin, Marcia P., and Kathryn V. Anderson. "A Conserved Signaling Pathway: The Drosophila Toll-Dorsal Pathway." *Annual Review of Cell and Developmental Biology* 12 (1996): 395–416.

Benjamin, Ruha. "A Lab of Their Own: Genomic Sovereignty as Postcolonial Policy." *Policy and Society* 28 (2009): 341–55.

Benn, P. A., and A. R. Chapman. "Practical and Ethical Considerations of Noninvasive Prenatal Diagnosis." *Journal of the American Medical Association* 301, no. 20 (2009): 2154–56.

Bertalanffy, Ludwig von. *General Systems Theory: Foundations, Development, Applications*. New York: George Braziller, 1968.

Bertalanffy, Ludwig von. "An Outline for General Systems Theory." *British Journal for the Philosophy of Science* 1 (1950): 134–65.

Biggs, N., E. Lloyd, and R. Wilson. *Graph Theory, 1736–1936*. New York: Oxford University Press, 1986.

Billings, Paul R., Jonathan Beckwith, and Joseph S. Alper. "The Genetic Analysis of Human Behavior: A New Era?" *Social Science and Medicine* 35 (1992): 227–38.

"Biology 2.0." *Economist*. June 17, 2010. http://www.economist.com/node/163 -49358.

Black, Max. *Models and Metaphors: Studies in Language and Philosophy*. Ithaca, NY: Cornell University Press, 1962.

Blake, Judith A., and Carol J. Bult. "Beyond the Data Deluge: Data Integration and Bio-ontologies." *Journal of Biomedical Informatics* 39 (2005): 314–20.

Blanchard, R. "Fraternal Birth Order and the Maternal Immune Hypothesis of Male Homosexuality." *Hormones and Behavior* 40, no. 2 (2001): 105–14.

Blaxter, Mark. "Revealing the Dark Matter of the Genome." *Science* 330 (2010): 1758–59.

Bliss, Catherine. "Genome Sampling and the Biopolitics of Race." In *A Foucault for the 21st Century: Governmentality Biopolitics and Discipline in the New Millennium*, edited by Samuel Binkley and Jorge Capetilla, 322–29. Cambridge: Cambridge Scholars Publishing, 2009.

Bliss, Catherine. "The Marketization of Identity Politics." *Sociology* 47 (2013): 1011–25.

Bliss, Catherine. *Race Decoded: The Genomic Fight for Social Justice*. Stanford, CA: Stanford University Press, 2012.

Bliss, Catherine. "Translating Racial Genomics: Passages in and beyond the Lab." *Qualitative Sociology* 36 (2013): 423–43.

Blum, K., E. P. Noble, P. J. Sheridan, A. Montgomery, T. Ritchie, P. Jaga-deeswaran, H. Nogami, A. H. Briggs, and J. B. Cohn. "Allelic Association of Human Dopamine d2 Receptor Gene in Alcoholism." *Journal of the American Medical Association* 263 (1990): 2055–60.

Boardman, Jason D., Jonathan Daw, and Jeremy Freese. "Defining the Environment in Gene-Environment Research: Lessons from Social Epidemiology." *American Journal of Public Health*, e-View Ahead of Print, 2013.

Bocklandt, Sven, Steve Horvath, Eric Vilain, and Dean H. Hamer. "Extreme Skewing of X Chromosome Inactivation in Mothers of Homosexual Men." *Human Genetics* 118, no. 6 (2006): 691–94.

Börner, Katy. *Atlas of Science: Visualising What We Know*. Cambridge, MA: MIT Press, 2010.

Bosker, F. J., C. A. Hartman, I. M. Nolte, B. P. Prins, P. Terpstra, D. Posthuma, T. van Veen, et al. "Poor Replication of Candidate Genes for Major Depressive Disorder Using Genome-Wide Association Data." *Molecular Psychiatry* 16 (2011): 516–32.

Bouchard, Thomas J. "Genetic Influence on Human Intelligence (Spearman's G): How Much?" *Annals of Human Biology* 36 (2009): 527–44.

Bourdieu, Pierre. *Pascalian Meditations*. Translated by Richard Nice. Stanford, CA: Stanford University Press, 2000.

Bourdieu, Pierre. *Science of Science and Reflexivity*. Translated by Richard Nice. Chicago: University of Chicago Press, 2004.

Bowcock, Anne, and Luca Cavalli-Sforza. "The Study of Variation in the Human Genome." *Genomics* 11, no. 2 (1991): 491–98.

Bowker, Geoff C. *Memory Practices in the Sciences*. Cambridge, MA: MIT Press, 2006.

Brent, Roger. "After the Genome IV: Envisioning Biology in the Year 2010." NASA Technical Documents Archive, May 13, 1999. http://archive.org/details/nasa_techdoc_19990058170.

Brown, Frank. "Editorial Opinion: Chemoinformatics—a Ten Year Update." *Current Opinion in Drug Discovery and Development* 8, no. 3 (2005): 298–302.

Brulle, Robert J., and David N. Pellow. "Environmental Justice: Human Health and Environmental Inequalities." *Annual Review of Public Health* 27 (2006): 103–24.

Buchanan, Mark. *Nexus: Small Worlds and the Groundbreaking Science of Networks*. New York: W. W. Norton, 2002.

Buerkle, Tom. "The 'Wondrous Map' of Gene Data: Historic Moment for Humanity's Blueprint." *New York Times*, June 27, 2000. http://www.nytimes.com/2000/06/27/news/27iht-genome.2.t_0.html.

Buetow, Kenneth H. "Cyberinfrastructure: Empowering a 'Third Way' in Biomedical Research." *Science* 308 (2005): 821–24.

Burchard, Esteban G., E. Ziv, N. Coyle, S. L. Gomez, H. Tang, A. J. Karter, J. L. Mountain, E. J. Pérez-Stable, D. Sheppard, and Neil Risch. "The Importance of Race and Ethnic Background in Biomedical Research and Clinical Practice." *New England Journal of Medicine* 348 (2003): 1170–75.

Burks, Christian, Michael J. Cinkosky, Paul Gilna, Jamie E.-D. Hayden, Yuki Abe, Edwin J. Atencio, Steve Barnhouse, et al. "GenBank: Current Status and Future Directions." *Methods in Enzymology* 183 (1990): 3–22.

Burton, Paul R., David G. Clayton, Lon R. Cardon, Nick Craddock, Panos Deloukas, Audrey Duncanson, Dominic P. Kwiatkowski, et al. "Genome-Wide Association Study of 14,000 Cases of Seven Common Diseases and 3,000 Shared Controls." *Nature* 447, no. 7145 (2007): 661–78.

Burwell, R. Geoffrey, Peter H. Dangerfield, Alan Moulton, and Theodoros B Grivas. "Adolescent Idiopathic Scoliosis (AIS), Environment, Exposome and Epigenetics." *Scoliosis* 6 (2011): 26–46.

Bustamante, Carlos D., Francisco M. De La Vega, and Esteban G. Burchard. "Genomics for the World." *Nature* 475 (2011): 163–65.

Buxbaum, J. D., S. Baron-Cohen, and B. Devlin. "Genetics in Psychiatry: Common Variant Association Studies." *Molecular Autism* 1 (2010): 6.

Calhoun, Craig J. *Critical Social Theory: Culture, History, and the Challenge of Difference.* Oxford: Blackwell, 1995.

Callaway, Ewen. "Economics and Genetics Meet in Uneasy Union." *Nature* 490 (2012): 154–55.

Callebaut, Werner. "Scientific Perspectivism: A Philosopher of Science's Response to the Challenge of Big Data Biology." *Studies in the History and the Philosophy of the Biological and Biomedical Sciences* 43 (2012): 69–80.

Cantor, Rita M., Kenneth Lange, and Janet S. Sinsheimer. "Prioritizing GWAS Results: A Review of Statistical Methods and Recommendations for Their Application." *American Journal of Human Genetics* 86, no. 1 (2010): 6–22.

Carey, Nessa. *The Epigenetics Revolution: How Modern Biology Is Rewriting Our Understanding of Genetics, Disease, and Inheritance.* New York: Columbia University Press, 2012.

Carter-Pokras, Olivia, and Claudia Baquet. "What Is a 'Health Disparity'?" *Public Health Reports* 117 (2002): 426–34.

Casper, Monica J. *The Making of the Unborn Patient: A Social Anatomy of Fetal Surgery.* New Brunswick, NJ: Rutgers University Press, 1998.

Caspi, Avshalom, J. McClay, Terrie E. Moffitt, Jonathan Mill, Judy Martin, Ian W. Craig, Alan Taylor, and Richie Poulton. "Role of Genotype in the Cycle of Violence in Maltreated Children." *Science* 297 (2002): 851–54.

Caspi, Avshalom, Karen Sugden, Terrie E. Moffitt, Alan Taylor, Ian W. Craig, HonaLee Harrington, J. McClay, et al. "Influence of Life Stress on Depression: Moderation by a Polymorphism in the 5-Htt Gene." *Science* 301 (2003): 386–89.

Casselman, Anne. "Strange but True: The Largest Organism on Earth Is a Fungus." *Scientific American*, October 4, 2007. http://www.scientificamerican.com /article.cfm?id=strange-but-true-largest-organism-is-fungus.

Castells, Manuel. *The Internet Galaxy.* New York: Oxford University Press, 2001.

Centers for Disease Control and Prevention. "Public Health Genomics." 2011. http://www.cdc.gov/genomics/.

Centers for Disease Control and Prevention. "Use of Race and Ethnicity in Public Health Surveillance Summary of the CDC/ATSDR Workshop." 1993.

Centerwall, W. R., and K. Benirschke. "Male Tortoiseshell and Calico (T—C) Cats." *Journal of Heredity* 64 (1973): 272–78.

Champagne, F. A., and M. J. Meaney. "Stress during Gestation Alters Postpartum Maternal Care and the Development of the Offspring in a Rodent Model." *Biological Psychiatry* 59 (2006): 1227–35.

Chandler, David L. "Heredity Study Eyes European Origins." *Boston Globe*, May 10, 2001. http://www.highbeam.com/doc/1P2-8655716.html.

Charney, Evan. "Behavior Genetics and Postgenomics." *Behavioral and Brain Sciences* 35 (2012): 1–80.

Charney, Evan, and William English. "Candidate Genes and Political Behavior." *American Political Science Review* 106 (2012): 1–34.

Check Hayden, Erika. "Human Genome at Ten: Life Is Complicated." *Nature News* 464, no. 7289 (2010): 664–67.

Chen, Xi, and Hemant Ishwaran. "Random Forests for Genomic Data Analysis." *Genomics* 99, no. 6 (2012): 323–29.

Children's Environmental Health Network. "The Contribution of Epigenetics in Pediatric Environmental Health." *EpiGenie*, 2012. http://epigenie.com/conferences/the-contribution-of-epigenetics-in-pediatric-environmental-health/.

Christakis, Nicholas A., and James H. Fowler. "The Spread of Obesity in a Large Social Network over 32 Years." *New England Journal of Medicine* 357 (2007): 370–79.

Chuang, Han-Yu, Eunjung Lee, Yu-Tsueng Liu, Doheon Lee, and Trey Ideker. "Network-Based Classification of Breast Cancer Metastasis." *Molecular Systems Biology* 3 (2007): 1–10.

Church, Dawson. *The Genie in Your Genes: Epigenetic Medicine and the New Biology of Intention.* Santa Rosa, CA: Elite Books, 2007.

Cinkowky, Michael J., J. W. Fickett, Paul Gilna, and Christian Burks. "Electronic Data Publishing and GenBank." *Science* 252 (1991): 1273–77.

Clarke, Adele E., Janet Shim, Sara Shostak, and Alondra Nelson. "Biomedicalizing Genetic Health, Diseases and Identities." In *Handbook of Genetics and Society: Mapping the New Genomic Era*, edited by Paul Atkinson, Peter Glasner, and Margaret Lock. London: Routledge, 2009.

Cloud, John. "Why Your DNA Isn't Your Destiny." *Time*, January 6, 2010.

Cochran, Gregory, Jason Hardy, and Henry Harpending. "Natural History of Ashkenazi Intelligence." *Journal of Biosocial Science* 38 (2006): 659–93.

Cole, Steve W. "Social Regulation of Human Gene Expression." *Current Directions in Psychological Science* 18, no. 3 (2009): 132–37.

Cole, Steve W., Louise C. Hawkley, Jesusa M. Arevalo, Caroline Y. Sung, Robert M. Rose, and John T. Cacioppo. "Social Regulation of Gene Expression in Human Leukocytes." *Genome Biology* 8, no. 9 (2007): R189.

Collins, Francis S. "Contemplating the End of the Beginning." *Genome Research* 1 (2001): 641–43.

Collins, Francis S. "Has the Revolution Arrived?" *Nature* 464, no. 7289 (2010): 674–75.

Collins, Francis S. "Medical and Societal Consequences of the Human Genome Project." *New England Journal of Medicine* 341 (1999): 28–37.

Collins, Francis S., Lisa D. Brooks, and Aravinda Chakravarti. "A DNA Polymorphism Discovery Resource for Research on Human Genetic Variation." *Genome Research* 8 (1998): 1229–31.

Collins, Francis S., and D. Galas. "A New Five-Year Plan for the U.S. Human Genome Project." *Science* 262 (1993): 43–46.

Collins, Francis S., Eric D. Green, Alan E. Guttmacher, and Mark S. Guyer. "A Vision for the Future of Genomics Research." *Nature* 422 (2003): 835–47.

Committee on Network Science for Future Army Applications, National Research Council. *Network Science*. Washington, DC: National Academies Press, 2005.

Conrad, Peter, and Dana Weinberg. "Has the Gene for Alcoholism Been Discovered Three Times since 1980? A News Media Analysis." *Perspectives on Social Problems* 8 (1996): 3–25.

Contreras, Jorge L. "Bermuda's Legacy: Policy, Patents and the Genome Commons." *Minnesota Journal of Law, Science and Technology* 12 (2010): 61–102.

Cooney, Elizabeth. "Two-Way Exchange." *Broad Institute* (blog), June 20, 2012. http://www.broadinstitute.org/blog/two-way-exchange.

Corrigan, Oonagh, and Richard Tutton. "Biobanks and the Challenges of Governance, Legitimacy and Benefit." In *The Handbook of Genetics and Society: Mapping the New Genomic Era*, edited by Paul Atkinson, Peter E. Glasner, and Margaret M. Lock. New York: Routledge, 2009.

Coyne, Jerry. "Is 'Epigenetics' a Revolution in Evolution?" *Why Evolution Is True* (blog), August 21, 2011. http://whyevolutionistrue.wordpress.com/2011/08/21/is-epigenetics-a-revolution-in-evolution/.

Cronin, Bernard. *The Scholar's Courtesy: The Role of Acknowledgement in the Primary Communication Process*. London: Taylor Graham, 1995.

Cummins, Steven. "Improving Population Health through Area-Based Social Interventions: Generating Evidence in a Complex World." In *Evidence-Based Public Health: Effectiveness and Efficiency*, edited by Amanda Killoran and Mike P. Kelly, 287–97. New York: Oxford University Press, 2009.

Cummins, Steven. "Neighbourhood Food Environment And Diet—Time for Improved Conceptual Models?" *Preventive Medicine* 44 (2007): 196–97.

Cummins, Steven, Sarah Curtis, Ana V. Diez-Roux, and Sally Macintyre. "Understanding and Representing 'Place' in Health Research: A Relational Approach." *Social Science and Medicine* 65 (2007): 1825–38.

Cummins, Steven, Sally Macintyre, Sharon Davidson, and Anne Ellaway. "Measuring Neighbourhood Social and Material Context: Generation and Interpretation of Ecological Data from Routine and Non-routine Sources." *Health and Place* 11 (2005): 249–60.

Curry, Andrew. "Rescue of Old Data Offers Lesson for Particle Physicists." *Science* 331 (2011): 694–95.

Daniels, Cynthia R. "Between Fathers and Fetuses: The Social Construction of Male Reproduction and the Politics of Fetal Harm." *Signs* 22, no. 3 (1997): 579–616.

Daniels, Cynthia R. *Exposing Men: The Science and Politics of Male Reproduction.* New York: Oxford University Press, 2006.

Dardel, Frédéric, and François Képès. *Bioinformatics: Genomics and Post-genomics.* Hoboken, NJ: John Wiley and Sons, 2006.

Daston, Lorraine J., and Peter L. Galison. *Objectivity.* 1st ed. Cambridge, MA: Zone Books, 2007.

Daston, Lorraine J., and Peter L. Galison. *Objectivity.* 2nd ed. New York: Zone Books, 2010.

Davies, Gail, Emma Frow, and Sabina Leonelli. "Bigger, Faster, Better: Rhetorics and Practices of Large-Scale Research in Contemporary Bioscience." *BioSocieties* 8 (2013): 386–96.

Davies, Gail, Albert Tenesa, Antony Payton, Jian Yang, Sarah E. Harris, David Liewald, Xiayi Ke, et al. "Genome-Wide Association Studies Establish That Human Intelligence Is Highly Heritable and Polygenic." *Molecular Psychiatry* 16, no. 10 (2011): 996–1005.

Davis, Bernard D. "The Human Genome and Other Initiatives." *Science* 249 (1990): 342–43.

Davis, Nicole. "Mexico-US Collaboration Launched." *Broad Institute Communications,* January 19, 2010. http://www.broadinstitute.org/news/1405.

Dawkins, Richard. *The Blind Watchmaker.* New York: Norton, 1986.

Dawkins, Richard. *The Selfish Gene.* Oxford: Oxford University Press, 1976.

Day, Nathan, Andrew Hemmaplardh, Robert E. Thurman, John A. Stamatoyannopoulos, and William S. Noble. "Unsupervised Segmentation of Continuous Genomic Data." *Bioinformatics* 23, no. 11 (2007): 1424–26.

Deary, Ian J., W. Johnson, and L. M. Houlihan. "Genetic Foundations of Human Intelligence." *Human Genetics* 126 (2009): 215–32.

Deleuze, Gilles, and Felix Guattari. *What Is Philosophy? European Perspectives.* New York: Columbia University Press, 1994.

Derrida, Jacques. *Of Grammatology.* Translated by Gayatri Spivak. Baltimore: Johns Hopkins University Press, 1974.

Devitt, M. "Resurrecting Biological Essentialism." *Philosophy of Science* 75 (2008): 344–82.

De Vries, Jantina, Melodie Slabbert, and Michael S. Pepper. "Ethical, Legal and Social Issues in the Context of the Planning Stages of the Southern African Human Genome Programme." *Medicine and Law* 31 (2012): 119–52.

Dial, S., A. Kezouh, A. Dascal, A. Barkun, and S. Suissa. "Patterns of Antibiotic Use and Risk of Hospital Admission because of *Clostridium difficile* Infection." *Canadian Medical Association Journal* 179, no. 8 (2008): 767–72.

Dick, D. M., R. J. Rose, R. J. Viken, J. Kaprio, and M. Koskenvuo. "Exploring Gene-Environment Interactions: Socioregional Moderation of Alcohol Use." *Journal of Abnormal Psychology* 110 (2001): 625–32.

Dick, Danielle M., and Richard J. Rose. "Behavior Genetics: What's New? What's Next?" *Current Directions in Psychological Science* 11 (2002): 70–74.

Diez Roux, Ana V. "Bringing Context Back into Epidemiology: Variables and Fallacies in Multilevel Analysis." *American Journal of Public Health* 88 (1998): 216–22.

Diez Roux, Ana V. "Complex Systems Thinking and Current Impasses in Health Disparities Research." *American Journal of Public Health* 101 (2011): 1627–34.

Diez Roux, Ana V. "Neighborhoods and Health: Where Are We and Where Do We Go from Here?" *Revue d'Épidémiologie Et De Santé Publique* 55 (2007): 13–21.

Diez Roux, Ana V., and Christina Mair. "Neighborhoods and Health." *Annals of the New York Academy of Sciences* 1186 (2010): 125–45.

Diez Roux, Ana V., Sharon Stein Merkin, Donna Arnett, Lloyd Chambless, Mark Massing, F. Javier Nieto, Paul Sorlie, Moyses Szklo, Herman A. Tyroler, and Robert L. Watson. "Neighborhood of Residence and Incidence of Coronary Heart Disease." *New England Journal of Medicine* 345 (2001): 99–106.

Doolittle, W. Ford. "Is Junk DNA Bunk? A Critique of ENCODE." *Proceedings of the National Academy of Sciences* 110, no. 14 (2013): 5294–300.

Doolittle, W. Ford, and Carmen Sapienza. "Selfish Genes, the Phenotype Paradigm and Genome Evolution." *Nature* 284, no. 5757 (1980): 601–3.

Dorogovtsev, S. N., and J. F. F. Mendes. "Evolution of Networks." *Advances in Physics* 51 (2002): 1079–187.

Dupré, John. *Humans and Other Animals.* Oxford: Oxford University Press, 2002.

Dupré, John, and Maureen A. O'Malley. "Varieties of Living Things: Life at the Intersection of Lineage and Metabolism." *Philosophy and Theory in Biology* 1 (2009). doi.org/10.3998/ptb.6959004.0001.003.

Duster, Troy. *Backdoor to Eugenics.* New York: Routledge, 1990.

Duster, Troy. "Race and Reification in Science." *Science* 307 (2005): 1050–51.

Duster, Troy. "Selective Arrests, an Ever-Expanding DNA Forensic Database, and the Specter of an Early-Twenty-First-Century Equivalent of Phrenology." In *DNA and the Criminal Justice System: The Technology of Justice*, edited by D. Lazer, 315–34. Cambridge, MA: MIT Press, 2004.

Dyer, Jennifer Shine, and Charles R. Rosenfeld. "Metabolic Imprinting by Prenatal, Perinatal, and Postnatal Overnutrition: A Review." *Seminars in Reproductive Medicine* 29, no. 3 (2011): 266–76.

Edwards, O. M. "Masculinized Turner's Syndrome XY-XO Mosaicism." *Proceedings of the Royal Society of Medicine* 64, no. 3 (1971): 300–301.

Edwards, Paul. *A Vast Machine.* Cambridge, MA: MIT Press, 2010.

Edwards, Paul N., Matthew S. Mayemik, Archer L. Batcheller, Geoffrey C. Bowker, and Christine L. Borgman. "Science Friction: Data, Metadata, and Collaboration." *Social Studies of Science* 41, no. 5 (2011): 667–90.

Egeland, Janice A., Daniela S. Gerhard, David L. Pauls, James N. Sussex, Kenneth K. Kidd, Cleona R. Alien, Abram M. Hostetter, and David E. Housman. "Bipolar Affective Disorders Linked to DNA Markers on Chromosome 11." *Nature* 325 (1987): 783–87.

Ellen, Ingrid Gould, Tod Mijanovich, and Keri-Nicole Dillman. "Neighborhood Effects on Health: Exploring the Links and Assessing the Evidence." *Journal of Urban Affairs* 23 (2001): 391–408.

Ellis, Brian David. *Scientific Essentialism*. Cambridge: Cambridge University Press, 2001.

Ellis, James K., Toby J. Athersuch, Laura D. K. Thomas, Friedrike Teichert, Miriam Pérez-Trujillo, Claus Svendsen, David J. Spurgeon, et al. "Metabolic Profiling Detects Early Effects of Environmental and Lifestyle Exposure to Cadmium in a Human Population." *BMC Medicine* 10 (2012): 61–71.

The ENCODE Project Consortium. "The ENCODE (ENCyclopedia Of DNA Elements) Project." *Science* 306, no. 5696 (2004): 636–40.

The ENCODE Project Consortium. "Identification and Analysis of Functional Elements in 1% of the Human Genome by the ENCODE Pilot Project." *Nature* 447, no. 7146 (2007): 799–816.

The ENCODE Project Consortium. "An Integrated Encyclopedia of DNA Elements in the Human Genome." *Nature* 489, no. 7414 (2012): 57–74.

Epstein, Steven. "Bodily Differences and Collective Identities: The Politics of Gender and Race in Biomedical Research in the United States." *Body and Society* 10 (2004): 183–203.

Epstein, Steven. *Inclusion: The Politics of Difference in Medical Research*. Chicago: University of Chicago Press, 2007.

Erdős, Paul, and Alfréd Rényi. "On Random Graphs I." *Publicationes Mathematicae* 6 (1959): 290–97.

Evans, James, and Rzhesky, Andrew. "Machine Science." *Science* 329 (2010): 399–400.

Faisandier, Laurie, Vincent Bonneterre, Régis De Gaudemaris, and Dominique J. Bicout. "Occupational Exposome: A Network-Based Approach for Characterizing Occupational Health Problems." *Journal of Biomedical Informatics* 44 (2011): 545–52.

Feil, Robert, and Mario F. Fraga. "Epigenetics and the Environment: Emerging Patterns and Implications." *Nature Reviews Genetics* 13 (2012): 97–109.

Feinberg, Andrew P. "Epigenetics at the Epicenter of Modern Medicine." *Journal of the American Medical Association* 299, no. 11 (2008): 1345–50.

Fisher, R. A. "The Use of Multiple Measurements in Taxonomic Problems." *Annals of Human Genetics* 7, no. 2 (1936): 179–88.

Foerster, Heinz von. "A Predictive Model for Self-Organizing Systems (Part I)." *Cybernetica* 3 (1961): 258–300.

Fortun, Michael. "For an Ethics of Promising, or, a Few Kind Words about James Watson." *New Genetics and Society* 24, no. 2 (2005): 157–73.

Fortun, Michael. "The Human Genome Project: Past, Present, and Future Anterior." In *Science, History and Social Activism: A Tribute to Everett Mendelsohn*, edited by Garland E. Allen and Roy M. MacLeod, 339–62. Dordrecht: Kluwer, 2002.

Fortun, Michael. "Mapping and Making Genes and Histories: The Genomics Project in the US, 1980–1990." PhD diss., Harvard University, 1993.

Fortun, Michael. *Promising Genomics: Iceland and deCODE Genetics in a World of Speculation.* Berkeley: University of California Press, 2008.

Fortun, Michael, and Everett Mendelsohn. *The Practices of Human Genetics.* Boston: Kluwer, 1999.

Foucault, Michel. *The Essential Foucault: Selections from Essential Works of Foucault, 1954–1984.* Edited by Paul Rabinow and Nikolas Rose. New York: New Press, 2003.

Foucault, Michel. *The History of Sexuality.* Vol. 1. New York: Random House, 1978.

Francis, Richard C. *Epigenetics: The Ultimate Mystery of Inheritance.* New York: W. W. Norton, 2011.

Fry, Ben. *Visualising Data: Explaining and Exploring Data with the Processing Environment.* Beijing: O'Reilly, 2008.

Fujimura, Joan H. "Postgenomic Futures: Translations across the Machine-Nature Border in Systems Biology." *New Genetics and Society* 24, no. 2 (2005): 195–226.

Fujimura, Joan H., and Ramya Rajagopalan. "Different Differences: The Use of 'Genetic Ancestry' versus Race in Biomedical Human Genetic Research." *Social Studies of Science* 41, no. 5 (2011): 5–30.

Galison, Peter L. "The Ontology of the Enemy: Norbert Wiener and the Cybernetic Vision." *Critical Inquiry* 21 (1994): 228–66.

Garcia-Sancho, Miguel. *Biology, Computing, and the History of Molecular Sequencing: From Proteins to DNA, 1945–2000.* New York: Palgrave Macmillan, 2012.

Gatherer, Derek. "So What Do We Really Mean When We Say That Systems Biology Is Holistic?" *BMC Systems Biology* 4 (2010): 22.

Gee, Gilbert C., and Devon C. Payne-Sturges. "Environmental Health Disparities: A Framework Integrating Psychosocial and Environmental Concepts." *Environmental Health Perspectives* 112 (2004): 1645–53.

GENEVA. "GENEVA Study Overview." 2011. https://www.genevastudy.org/Study Overview.

"The Genome, 10 Years Later." *New York Times*, June 20, 2010.

Gerovich, Slava. *From Newspeak to Cyberspeak: A History of Soviet Cybernetics.* Cambridge, MA: MIT Press, 2002.

Ghose, Tia. "Heritability of Intelligence." *Scientist*, August 9, 2011. http://the -scientist.com/2011/08/09/heritability-of-intelligence/.

Gibbs, W. Wayt. "The Unseen Genome: Gems among the Junk." *Scientific American*, November 2003, 46–53.

Gieryn, Thomas F. "A Space for Place in Sociology." *Annual Review of Sociology* 26 (2000): 463.

Gilbert, Walter. "Towards a Paradigm Shift in Biology." *Nature* 349 (1991): 99.

Gilbert, Walter. "A Vision of the Grail." In *The Code of Codes: Scientific and Social Issues in the Human Genome Project*, edited by D. J. Kevles and L. E. Hood, 83–97. Cambridge, MA: Harvard University Press, 1992.

Giordano, Antonio, and Nicola Normanno. *Breast Cancer in the Post-genomic Era.* New York: Humana Press, 2009.

Gissis, Snait, and Eva Jablonka. *Transformations of Lamarckism: From Subtle Fluids to Molecular Biology.* Cambridge, MA: MIT Press, 2011.

Gladwell, Malcolm. *The Tipping Point: How Little Things Can Make a Big Difference.* Boston: Little, Brown, 2000.

Gluckman, Peter D., and Mark A. Hanson. *Developmental Origins of Health and Disease.* New York: Cambridge University Press, 2006.

Gluckman, Peter D., and Mark A. Hanson. *The Fetal Matrix: Evolution, Development, and Disease.* New York: Cambridge University Press, 2005.

Godfrey-Smith, Peter. "Information in Biology." In *The Cambridge Companion to the Philosophy of Biology,* edited by D. Hull and M. Ruse, 103–19. Cambridge: Cambridge University Press, 2007.

Google Inc. "Behind the Compute Engine Demo at Google I/O 2012 Keynote." 2012. https://developers.google.com/compute/io.

Gorman, James. "Mouse Study Discovers DNA That Controls Behavior." *New York Times,* January 16, 2013.

Gorroochurn, Prakash, Susan E. Hodge, Gary A. Heiman, Martina Durner, and David A. Greenberg. "Non-replication of Association Studies: 'Pseudo-failures' to Replicate?" *Genetics in Medicine* 9 (2007): 325–31.

Gottesman, Irving I., and Todd D. Gould. "The Endophenotype Concept in Psychiatry: Etymology and Strategic Intentions." *American Journal of Psychiatry* 106 (2003): 636–45.

Gottlieb, G. "Some Conceptual Deficiencies in 'Developmental' Behavior Genetics." *Human Development* 38 (1995): 131–41.

Granovetter, Mark. "The Strength of Weak Ties." *American Journal of Sociology* 78 (1973): 1360–80.

Graur, Dan, Yichen Zheng, Nicholas Price, Ricardo B. R. Azevedo, Rebecca A. Zufall, and Eran Elhaik. "On the Immortality of Television Sets: 'Function' in the Human Genome according to the Evolution-Free Gospel of ENCODE." *Genome Biology and Evolution* (2013).

Greally, John M. "Genomics: Encyclopaedia of Humble DNA." *Nature* 447 (2007): 782–83.

Green, Eric D., Mark S. Guyer, and National Human Genome Research Institute. "Charting a Course for Genomic Medicine from Base Pairs to Bedside." *Nature* 470 (2011): 204–13.

Griesemer, James. "Turning Back to Go Forward." *Biology and Philosophy* 13, no. 1 (1998): 103–12.

Griffiths, P. "Genetic Information: A Metaphor in Search of a Theory." *Philosophy of Science* 68 (2001): 394–412.

Gross, Michael. "African Genomes." *Current Biology* 21 (2011): R481–R484.

Gutierrez, Alberto. "Letter to Anne Wojcicki. 23andMe, Inc. 11/22/13." U.S. Food and Drug Administration, November 22, 2013. http://www.fda.gov/ICECI/EnforcementActions/WarningLetters/2013/ucm376296.htm.

Guttmacher, Alan E., and Francis S. Collins. "Realizing the Promise of Genomics in Biomedical Research." *Journal of the American Medical Association* 294 (2005): 1399–402.

Guttmacher, Alan E., and Francis S. Collins. "Welcome to the Genomic Era." *New England Journal of Medicine* 349 (2003): 996–98.

Haig, D. "Weismann Rules! OK? Epigenetics and the Lamarckian Temptation." *Biology and Philosophy* 22, no. 3 (2007): 415–28.

Hall, P. A. "Policy Paradigms, Social Learning, and the State: The Case of Economic Policymaking in Britain." *Comparative Politics* 25 (1993): 275–96.

Hall, Stephen S. "Small and Thin: The Controversy over the Fetal Origins of Adult Health." *New Yorker*, November 19, 2007, 52–57.

Hamer, Dean H., and Peter Copeland. *The Science of Desire: The Search for the Gay Gene and the Biology of Behavior.* New York: Simon and Schuster, 1994.

Hamer, Dean H., Stella Hu, Victoria L. Magnuson, Nan Hu, and Angela M. L. Pattatucci. "A Linkage between DNA Markers on the X Chromosome and Male Sexual Orientation." *Science* 261 (1993): 321–25.

Hansson, Göran K., and Kristina Edfeldt. "Toll to Be Paid at the Gateway to the Vessel Wall." *Arteriosclerosis, Thrombosis, and Vascular Biology* 25, no. 6 (2005): 1085–87.

Haraway, Donna. *Crystals, Fabrics, and Fields: Metaphors of Organicism in Twentieth-Century Developmental Biology.* New Haven, CT: Yale University Press, 1976.

Harmon, Katherine. "Sequencing the 'Exposome': Researchers Take a Cue from Genomics to Decipher Environmental Exposure's Links to Disease." *Scientific American*, October 21, 2010. http://www.scientificamerican.com/article.cfm?id=environmental-exposure.

Harper, J. L. *Population Biology of Plants.* London: Academic Press, 1977.

Hart, M. H., R. J. Reader, and J. N. Klironomos. "Plant Coexistence Mediated by Arbuscular Mycorrhizal Fungi." *Trends in Ecology and Evolution* 18 (2003): 418–23.

Hastie, Trevor, Robert Tibshirani, and Jerome H. Friedman. *The Elements of Statistical Learning: Data Mining, Inference, and Prediction.* New York: Springer, 2009.

Hatemi, Peter K., and Rose McDermott. "The Genetics of Politics: Discovery, Challenges, and Progress." *Trends in Genetics* 28, no. 10 (2012): 525–33.

Heims, Steve J. *Constructing a Social Science for Postwar America: The Cybernetics Group 1945–1953.* Cambridge, MA: MIT Press, 1993.

Helmreich, Stefan. "Trees and Seas of Information: Alien Kinship and the Biopolitics of Gene Transfer in Marine Biology and Biotechnology." *American Ethnologist* 30 (2003): 340–58.

Henig, Robin Marantz. "The Genome in Black and White (and Gray)." *New York Times*, October 10, 2004. http://www.nytimes.com/2004/10/10/magazine/10GENETIC.html.

Hesse, Mary B. *Models and Analogies in Science.* London: Sheed and Ward, 1963.

Hey, Tony, Stewart Tansley, and Kristine Tolle, eds. *The Fourth Paradigm: Data-Intensive Scientific Discovery.* Redmond, WA: Microsoft Research, 2009.

Hilgartner, Stephen. "Biomolecular Databases: New Communication Regimes for Biology?" *Science Communication* 17 (1995): 240–63.

Hilgartner, Stephen. "Constituting Large-Scale Biology: Building a Regime of Governance in the Early Years of the Human Genome Project." *Biosocieties* 8, no. 4 (2013): 397–416.

Hilgartner, Stephen, ed. *Science and Democracy: Making Knowledge and Making Power in the Biosciences and Beyond.* London: Routledge, 2014.

Hilgartner, Stephen, and S. Brandt-Rauf. "Data Access, Ownership, and Control: Toward Empirical Studies of Access Practices." *Knowledge: Creation, Diffusion, Utilization* 15 (1994): 355–72.

Hine, Christine. "Databases as Scientific Instruments and Their Role in the Ordering of Scientific Work." *Social Studies of Science* 36 (2006): 269–98.

Hirsch, Jerry. "Behavior-Genetic, or 'Experimental,' Analysis: The Challenge of Science versus the Lure of Technology." *American Psychologist* 22 (1967): 118–30.

Hirschhorn, Joel N., Kirk Lohmuller, Edward Byrne, and Kurt Hirschhorn. "A Comprehensive Review of Genetic Association Studies." *Genetics in Medicine* 4 (2002): 45–61.

Hoffman, M. M., O. J. Buske, J. Wang, Z. Weng, J. A. Bilmes, and W. S. Noble. "Unsupervised Pattern Discovery in Human Chromatin Structure through Genomic Segmentation." *Nature Methods* 9, no. 5 (2012): 473–76.

Hoffman, Moshe, Uri Gneezy, and John A. List. "Nurture Affects Gender Differences in Spatial Abilities." *Proceedings of the National Academy of Sciences* 108, no. 36 (2011): 14786–88.

Holden, Constance. "A Cautionary Genetic Tale: The Sobering Story of D_2." *Science* 264 (1994): 1696–97.

Holland, John H. *Adaptation in Natural and Artificial Systems: An Introductory Analysis with Applications to Biology, Control, and Artificial Intelligence.* Cambridge, MA: MIT Press, 1992.

Holt, Sarah. "Epigenetics." In NOVA *ScienceNow.* Boston: WGBH, 2007.

Holt, Sarah, and Nigel Paterson. "Ghost in Your Genes." NOVA. Boston: WGBH, 2008.

Hood, Leroy. "Biology and Medicine in the Twenty-First Century." In *The Code of Codes: Scientific and Social Issues in the Human Genome Project*, edited by Daniel J. Kevles and Leroy Hood, 136–63. Cambridge, MA: Harvard University Press, 1992.

Hooshangi, Sara, Stephan Thiberge, and Ron Weiss. "Ultrasensitivity and Noise Propagation in a Synthetic Transcriptional Cascade." *Proceedings of the National Academy of Sciences USA* 102 (2005): 3581–86.

Howe, Doug, Maria Costanzo, Petra Fey, Takashi Gojobori, Linda Hannick, Winston Hide, David P. Hill, et al. "Big Data: The Future of Biocuration." *Nature* 455 (2008): 47–50.

Huala, Eva, Allan W. Dickerman, Margarita Garcia-Hernandez, Danforth Weems, Leonore Reiser, Frank LaFond, David Hanley, et al. "The Arabidopsis Information Resource (TAIR): A Comprehensive Database and Web-Based Information Retrieval, Analysis, and Visualization System for a Model Plant." *Nucleic Acids Research* 29 (2001): 102–5.

The Human Genome Project: Hearing Before the Subcommittee on Energy Research and Development, Committee on Energy and Natural Resources, United States Senate. 101st Cong. 101–894. July 11, 1990.

Hunt-Grubbe, Charlotte. "The Elementary DNA of Dr Watson." *Sunday Times* (London), October 14, 2007.

Hutter, Carolyn M., Alicia M. Young, Heather M. Ochs-Balcom, Cara L. Carty, Tao Wang, Christina T. L. Chen, Thomas E. Rohan, Charles Kooperberg, and Ulrike Peters. "Replication of Breast Cancer GWAS Susceptibility Loci in the Women's Health Initiative African American SHARe Study." *Cancer Epidemiology Biomarkers and Prevention* 20, no. 9 (2011): 1950–59.

International Committee of Medical Journal Editors (ICMJE). "Uniform Requirements for Manuscripts Submitted to Biomedical Journals." *Annals of Internal Medicine* 126 (1997): 36–47.

The International HapMap Consortium. "Integrating Ethics and Science in the International HapMap Project." *Nature Reviews Genetics* 5 (2004): 467–75.

The International HapMap Consortium. "The International HapMap Project." *Nature* 426 (2003): 789–94.

Ito, Takashi, Tomoko Chiba, Ritsuko Ozawa, Mikio Yoshida, Masahira Hattori, and Yoshiyuki Sakaki. "A Comprehensive Two-Hybrid Analysis to Explore the Yeast Protein Interactome." *Proceedings of the National Academy of Sciences USA* 98 (2001): 4569–74.

Jablonka, Eva. "Lamarckian Problematics in Biology." In *Transformations of Lamarckism: From Subtle Fluids to Molecular Biology*, edited by Snait Gissis and Eva Jablonka. Cambridge, MA: MIT Press, 2011.

Jackson, Myles. *The Genealogy of a Gene: Patents, HIV/AIDS, and Race*. Cambridge, MA: MIT Press, forthcoming.

Jacob, Francois. *The Possible and the Actual*. Seattle: University of Washington Press, 1982.

Jaroff, Leon. "The Gene Hunt." *Time*, March 20, 1989, 62–67.

Jasanoff, Sheila. *Reframing Rights: Bioconstitutionalism in the Genetic Age*. Cambridge, MA: MIT Press, 2011.

Jasny, B. R., and L. B. Ray. "Life and Art of Networks." *Science* 301 (2003): 1863.

Jeong, H., B. Tombor, R. Albert, Z. Oltvai, and Albert-László Barabási. "The Large-Scale Organization of Metabolic Networks." *Nature* 407 (2000): 651–55.

Jerrett, Michael, and Murray Finkelstein. "Geographies of Risk in Studies Linking Chronic Air Pollution Exposure to Health Outcomes." *Journal of Toxicology and Environmental Health, Part A: Current Issues* 68 (2005): 1207–42.

Johannsen, Wilhelm. "Some Remarks about Units in Heredity." *Hereditas* 4 (1923): 133–41.

Jones, D. A. "What Does the British Public Think about Human-Animal Hybrid Embryos?" *Journal of Medical Ethics* 35 (2009): 168–70.

Jones, Dean P., Youngja Park, and Thomas R. Ziegler. "Nutritional Metabolomics: Progress in Addressing Complexity in Diet and Health." *Annual Review of Nutrition* 32 (2012): 183–202.

Joseph, Jay. "The Crumbling Pillars of Behavioral Genetics." *GeneWatch* 24 (2011): 4–7.

Joseph, Jay. *The Missing Gene: Psychiatry, Heredity, and the Fruitless Search for Genes.* New York: Algora, 2006.

Kahn, Jonathan D. "Beyond BiDil: The Expanding Embrace of Race in Biomedical Research and Product Development." *Saint Louis University Journal of Health, Law, and Public Policy* 3 (2009): 61–92.

Kaiser, Jocelyn. "The Genome Project: What Will It Do as a Teenager?" *Science* 331 (2011): 660.

Kang, Sun J., Charleston W. K. Chiang, Cameron D. Palmer, Bamidele O. Tayo, Guillaume Lettre, Johannah L. Butler, Rachel Hackett, et al. "Genome-Wide Association of Anthropometric Traits in African- and African-Derived Populations." *Human Molecular Genetics* 19 (2010): 2725–38.

Kaplan, Jonathan Michael. *The Limits and Lies of Human Genetic Research: Dangers for Social Policy.* New York: Routledge, 2000.

Kato, Tadafumi. "Molecular Genetics of Bipolar Disorder and Depression." *Psychiatry and Clinical Neurosciences* 61 (2007): 3–19.

Kay, Lily E. *Who Wrote the Book of Life? A History of the Genetic Code.* Stanford, CA: Stanford University Press, 2000.

Keating, Peter, and Alberto Cambrosio. "Too Many Numbers: Microarrays in Clinical Cancer Research." *Studies in History and Philosophy of Science Part C: Studies in History and Philosophy of Biological and Biomedical Sciences* 43, no. 1 (2012): 37–51.

Keiger, Dale. "Looking for the Next Big Thing." *Notre Dame Magazine*, 2007. http://magazine.nd.edu/news/9883-looking-for-the-next-big-thing/.

Kell, Douglas B., and Stephen G. Oliver. "Here Is the Evidence, Now What Is the Hypothesis? The Complementary Roles of Inductive and Hypothesis-Driven Science in the Post-genomic Era." *BioEssays* 26 (2004): 99–105.

Keller, Evelyn Fox. *The Century of the Gene.* Cambridge, MA: Harvard University Press, 2000.

Keller, Evelyn Fox. "Developmental Biology as a Feminist Cause?" *Osiris* 12 (1997): 16–28.

Keller, Evelyn Fox. "*Drosophila* Embryos as Transitional Objects." *Historical Studies in the Physical and Biological Sciences* 36, no. 2 (1996): 313–46.

Keller, Evelyn Fox. "Dynamic Objectivity: Love, Power, and Knowledge." In *Reflections on Gender and Science*, 115–26. New Haven: Yale University Press, 1985.

Keller, Evelyn Fox. *A Feeling for the Organism: The Life and Work of Barbara McClintock.* New York: W. H. Freeman, 1983.

Keller, Evelyn Fox. *Making Sense of Life: Explaining Biological Development with Models, Metaphors, and Machines.* Cambridge, MA: Harvard University Press, 2002.

Keller, Evelyn Fox. *The Mirage of a Space between Nature and Nurture.* Durham, NC: Duke University Press, 2010.

Keller, Evelyn Fox. *Refiguring Life: Metaphors of Twentieth Century Biology.* New York: Columbia University Press, 1995.

Keller, Evelyn Fox. "Revisiting 'Scale-Free' Networks." *BioEssays* 27 (2005): 1060–68.

Kelsoe, John R., Edward I. Ginns, Janice A. Egeland, Daniela S. Gerhard, Alisa M. Goldstein, Sherri J. Bale, David L. Pauls, et al. "Re-evaluation of the Linkage Relationship between Chromosome 11p Loci and the Gene for Bipolar Affective Disorder in the Old Order Amish." *Nature* 342 (1989): 238–43.

Kelty, Christopher M. "This Is Not an Article: Model Organism Newsletters and the Question of 'Open Science.'" *Biosocieties* 7 (2012): 140–68.

Kendler, Kenneth S. "Discussion: Genetic Analysis." In *Genetic Approaches to Mental Disorders*, edited by E. S. Gershon and C. R. Cloninger, 99–106. Washington, DC: American Psychiatric Press, 1994.

Kendler, Kenneth S. "Psychiatric Genetics: A Methodologic Critique." *American Journal of Psychiatry* 162 (2005): 3–11.

Kevles, Daniel J. *In the Name of Eugenics: Genetics and the Uses of Human Heredity.* New York: Knopf, 1985.

Kevles, Daniel J., and Leroy E. Hood, eds. *The Code of Codes: Scientific and Social Issues in the Human Genome Project.* Cambridge, MA: Harvard University Press, 1992.

Kittles, Rick. "Genes and Environments: Moving toward Personalized Medicine in the Context of Health Disparities." *Ethnicity and Disease* 22 (2012): S1-43-6.

Knight, Danielle. "Gene Project Deemed Unethical." *Inter Press Service News Agency*, October 28, 1997. http://www.ipsnews.net/1997/10/rights-science-gene-project-deemed-unethical/.

Koenig, Barbara A., Sandra Soo-Jin Lee, and Sarah S. Richardson, eds. *Revisiting Race in a Genomic Age.* New Brunswick, NJ: Rutgers University Press, 2008.

Kolata, Gina. "Scientists to Seek Clues to Violence in Genome of Gunman in Newtown, Conn." *New York Times*, December 24, 2012.

Koonin, E. V., Y. I. Wolf, and G. P. Karev. "The Structure of the Protein Universe and Genome Evolution." *Nature* 420 (2002): 218–23.

Koselleck, Reinhart. *The Practice of Conceptual History: Timing History, Spacing Concepts.* Stanford, CA: Stanford University Press, 2002.

Koshland, Daniel. "Nature, Nurture and Behavior." *Science* 235 (1987): 1445.

Koshland, Daniel. "A Rational Approach to the Irrational." *Science* 250 (1990): 189.

Koshland, Daniel. "Sequences and Consequences of the Human Genome." *Science* 246 (1989): 189.

Krieger, Nancy. "Embodiment: A Conceptual Glossary for Epidemiology." *Journal of Epidemiology and Community Health* 59 (2005): 350–55.

Krieger, Nancy. "Stormy Weather: Race, Gene Expression, and the Science of Health Disparities." *American Journal of Public Health* 95 (2005): 2155–60.

Krieger, Nancy. "Theories for Social Epidemiology in the 21st Century: An Ecosocial Perspective." *International Journal of Epidemiology* 30 (2001): 668–77.

Kripke, S. *Naming and Necessity.* Cambridge, MA: Harvard University Press, 1980.

Krzywinski, Martin I., Jacqueline E. Schein, Inanc Birol, Joseph Connors, Randy Gascoyne, Doug Horsman, Steven J. Jones, and Marco A. Marra. "Circos: An

Information Aesthetic for Comparative Genomics." *Genome Research* 19, no. 9 (2009): 1639–45.

Kukla, Rebecca. *Mass Hysteria: Medicine, Culture, and Mothers' Bodies*. Lanham, MD: Rowman and Littlefield, 2005.

Kuzawa, Christopher W. "Fetal Origins of Developmental Plasticity: Are Fetal Cues Reliable Predictors of Future Nutritional Environments?" *American Journal of Human Biology* 17, no. 1 (2005): 5–21.

Kuzawa, Christopher W. "Why Evolution Needs Development, and Medicine Needs Evolution." *International Journal of Epidemiology* 41, no. 1 (2012): 223–29.

Kuzawa, Christopher W., Peter D. Gluckman, and Mark A. Hanson. "Developmental Perspectives on the Origins of Obesity." In *Adipose Tissue and Adipokines in Health and Disease*, edited by E. Fantuzzi and T. Mazzone, 207–20. Totowa, NJ: Humana Press, 2007.

Kuzawa, Christopher W., and Elizabeth Sweet. "Epigenetics and the Embodiment of Race: Developmental Origins of US Racial Disparities in Cardiovascular Health." *American Journal of Human Biology* 21, no. 1 (2009): 2–15.

Ladd-Taylor, Molly, and Lauri Umansky. *"Bad" Mothers: The Politics of Blame in Twentieth-Century America*. New York: New York University Press, 1998.

Landecker, Hannah. "Food as Exposure: Nutritional Epigenetics and the New Metabolism." *BioSocieties* 6, no. 2 (2011): 167–94.

Lander, Eric S. "Initial Impact of the Sequencing of the Human Genome." *Nature* 470, no. 7333 (2011): 187–97.

Lango Allen, Hana, Karol Estrada, Guillaume Lettre, Sonja I. Berndt, Michael N. Weedon, Fernando Rivadeneira, Cristen J. Willer, et al. "Hundreds of Variants Clustered in Genomic Loci and Biological Pathways Affect Human Height." *Nature* 467, no. 7317 (2010): 832–38.

Larson, Christina. "Inside China's Genome Factory." *MIT Technology Review*, February 11, 2013. http://www.technologyreview.com/featuredstory/511051/inside-chinas-genome-factory/.

Latkin, Carl A., and Aaron D. Curry. "Stressful Neighborhoods and Depression: A Prospective Study of the Impact of Neighborhood Disorder." *Journal of Health and Social Behavior* 44 (2003): 34–44.

Leatherdale, W. H. *The Role of Analogy, Model, and Metaphor in Science*. New York: American Elsevier, 1974.

Lee, T. M., and I. Zucker. "Vole Infant Development Is Influenced Perinatally by Maternal Photoperiodic History." *American Journal of Physiology: Regulatory, Integrative and Comparative Physiology* 255, no. 5 (1988): R831–R838.

Lemaitre, Bruno. "The Road to Toll." *Nature Reviews Immunology* 4, no. 7 (2004): 521–27.

Lenoir, Timothy, ed. *Inscribing Science: Scientific Texts and the Materiality of Communication*. Stanford, CA: Stanford University Press, 1998.

Lenoir, Timothy. "Models and Instruments in the Development of Electrophysiology, 1845–1912." *Historical Studies in the Physical and Biological Sciences* 17 (1986): 1–54.

Leonelli, Sabina. "Centralising Labels to Distribute Data: The Regulatory Role of Genomic Consortia." In *The Handbook for Genetics and Society: Mapping the New Genomic Era*, edited by P. Atkinson, P. Glasner, and M. Lock, 469–85. London: Routledge, 2009.

Leonelli, Sabina. "Classificatory Theory in Data-Intensive Science: The Case of Open Biomedical Ontologies." *International Studies in the Philosophy of Science* 26 (2012): 47–65.

Leonelli, Sabina, ed. "Data-Driven Research in the Biological and Biomedical Sciences." Special section in *Studies in the History and Philosophy of Science Part C: Studies in the History and Philosophy of the Biological and Biomedical Sciences* 43 (2012): 1–87.

Leonelli, Sabina. "Integrating Data to Acquire New Knowledge: Three Modes of Integration in Plant Science." *Studies in History and Philosophy of Biological and Biomedical Sciences* (2013), Online First. doi:10.1016/j.shpsc.2013.03.020.

Leonelli, Sabina. "Making Sense of Data-Driven Research in the Biological and the Biomedical Sciences." *Studies in History and Philosophy of the Biological and Biomedical Sciences* 43 (2012): 1–3.

Leonelli, Sabina. "Packaging Data for Re-use: Databases in Model Organism Biology." In *How Well Do Facts Travel? The Dissemination of Reliable Knowledge*, edited by P. Howlett and M. S. Morgan, 325–48. Cambridge: Cambridge University Press, 2010.

Leonelli, Sabina. "When Humans Are the Exception: Cross-Species Databases at the Interface of Clinical and Biological Research." *Social Studies of Science* 42 (2012): 214–36.

Leonelli, Sabina. "Why the Current Insistence on Open Access to Scientific Data? Big Data, Knowledge Production and the Political Economy of Contemporary Biology." *Bulletin of Science, Technology and Society* 33 (2013): 6–11.

Leonelli, Sabina, and Rachel A. Ankeny. "Re-thinking Organisms: The Impact of Databases on Model Organism Biology." *Studies in History and Philosophy of Biological and Biomedical Sciences* 43 (2012): 29–36.

Leonelli, Sabina, Nicholas Smirnoff, Jonathan Moore, and Ruth Bastow. "Making Open Data Work in Plant Science." *Journal for Experimental Botany* (2013). Early online. doi:10.1093/jxb/ert273.

Leventhal, Tama, and Jeanne Brooks-Gunn. "Moving to Opportunity: An Experimental Study of Neighborhood Effects on Mental Health." *American Journal of Public Health* 93 (2003): 1576–82.

Lewin, Roger. "DNA Databases Are Swamped." *Science* 232 (1986): 1599.

Lewontin, Richard C. "The Analysis of Variance and the Analysis of Causes." *American Journal of Human Genetics* 26 (1974): 400–411.

Lewontin, Richard C., Steven P. R. Rose, and Leon J. Kamin. *Not in Our Genes: Biology, Ideology, and Human Nature*. New York: Pantheon Books, 1984.

Lo, Y. M. D. "Fetal DNA in Maternal Plasma: Biology and Diagnostic Applications." *Clinical Chemistry* 46 (2000): 1903–6.

Lock, Margaret. "Eclipse of the Gene and the Return of Divination." *Current An-thropology* 46, no. S5 (2005): S47–S70.

Lock, Margaret. "The Lure of the Epigenome." *Lancet* 381, no. 9881 (2013): 1896–97.

Longino, Helen E. *Science as Social Knowledge: Values and Objectivity in Scientific Inquiry.* Princeton, NJ: Princeton University Press, 1990.

Lupski, James R. "Genomic Disorders: Structural Features of the Genome Can Lead to DNA Rearrangements and Human Disease Traits." *Trends in Genetics* 14, no. 10 (1998): 417–22.

Lupski, James R. "Genomic Disorders Ten Years On." *Genome Med* 1, no. 4 (2009): 42.

Maamar, H., A. Raj, and D. Dubnau. "Noise in Gene Expression Determines Cell Fate in *Bacillus subtilis.*" *Science* 317 (2007): 526–29.

MacArthur, Daniel. "Why Do Genome-Wide Scans Fail?" *Wired Science Blogs*, September 15, 2008. http://www.wired.com/wiredscience/2008/09/why-do -genome-wide-scans-fail/.

MacIntyre, Sally, and Anne Ellaway. "Neighbourhoods and Health: An Overview." In *Neighborhoods and Health*, 1st ed., edited by Ichiro Kawachi and Lisa F. Berk-man. New York: Oxford University Press, 2003.

Maderspacher, Florian. "Lysenko Rising." *Current Biology* 20, no. 19 (2010): R835–R837.

Maggio, John, Leslie Asako Gladsjo, Sue Williams, et al. *Faces of America.* PBS, 2010.

Maher, Brendan. "Personal Genomes: The Case of the Missing Heritability." *Nature News* 456, no. 7218 (2008): 18–21.

Mariscal, Gonzalo, Oscar Marban, and Covadonga Fernandez. "A Survey of Data Mining and Knowledge Discovery Process Models and Methodologies." *Knowl-edge Engineering Review* 25 (2010): 137–66.

Marks, James. "Why Your Zipcode May Be More Important to Your Health Than Your Genetic Code." *Huffington Post* May 24, 2009 (updated May 25, 2011). http://www.huffingtonpost.com/james-s-marks/why-your-zip-code-may-be _b_190650.html.

Marshall, Eliot. "Highs and Lows on the Research Roller Coaster." *Science* 264 (1994): 1693–95.

Marshall Graves, J. A., and C. M. Disteche. "Does Gene Dosage Really Matter?" *Journal of Biology* 6, no. 1 (2007): 1.

Massey, Douglas, and Nancy Denton. *American Apartheid: Segregation and the Mak-ing of the Underclass.* Cambridge, MA: Harvard University Press, 1993.

Mathias, Rasika A., Audrey V. Grant, Nicholas Rafaels, Tracey Hand, Li Gao, Can-delaria Vergara, Yuhjung J. Tsai, et al. "A Genome-Wide Association Study on African-Ancestry Populations for Asthma." *Journal of Allergy and Clinical Im-munology* 125 (2010): 336–46.e4.

Mattick, John S. "The Central Role of RNA in Human Development and Cognition." *FEBS Letters* 585 (2011): 1600–16.

Mattick, John S. "RNA as the Substrate for Epigenome-Environment Interactions: RNA Guidance of Epigenetic Processes and the Expansion of RNA Editing in

Animals Underpins Development, Phenotypic Plasticity, Learning, and Cognition." *Bioessays* 32 (2010): 548–52.

Mattick, John S. "RNA Regulation: A New Genetics?" *Nature Reviews Genetics* 5 (2004): 316–23.

Mattick, John S., and Mark F. Mehler. "RNA Editing, DNA Recoding and the Evolution of Human Cognition." *Trends in Neuroscience* 31 (2008): 227–33.

Maturana, Humberto, Francisco Varela, and R. Uribe. "Autopoiesis: The Organization of Living Systems, Its Characterization and a Model." *Biosystems* 5 (1974): 187–96.

Maynard Smith, J. "The Concept of Information in Biology." *Philosophy of Science* 67 (2000): 177–94.

McCain, Katherine W. "Communication, Competition, and Secrecy: The Production and Dissemination of Research-Related Information in Genetics." *Science, Technology, and Human Values* 16 (1991): 491–516.

McCain, Katherine W. "Mandating Sharing: Journal Policies in the Natural Sciences." *Science Communication* 16 (1995): 403–31.

McClellan, Jon, and Mary-Claire King. "Genetic Heterogeneity in Human Disease." *Cell* 141, no. 2 (2010): 210–17.

McDade, Lucinda A., David R. Maddison, Robert Guralnick, Heather A. Piwowar, Mary Liz Jameson, Kristofer M. Helgen, Patrick S. Herendeen, Andrew Hill, and Morgan L. Vis. "Biology Needs a Modern Assessment System for Professional Productivity." *BioScience* 61 (2011): 619–25.

McGowan, Patrick O., Aya Sasaki, Ana C. D'Alessio, Sergiy Dymov, Benoit Labonté, Moshe Szyf, Gustavo Turecki, and Michael J. Meaney. "Epigenetic Regulation of the Glucocorticoid Receptor in Human Brain Associates with Childhood Abuse." *Nature Neuroscience* 12, no. 3 (2009): 342–48.

McKinnon, Robin A., Jill Reedy, Meredith A. Morrissette, Leslie A. Lytle, and Amy L. Yaroch. "Measures of the Food Environment: A Compilation of the Literature, 1990–2007." *American Journal of Preventive Medicine* 36 (2009): S124–S133.

McKusick, Victor. "*Mendelian Inheritance in Man* and Its Online Version, OMIM." *American Journal of Human Genetics* 80 (2007): 588–604.

McNaughton, Darlene. "From the Womb to the Tomb: Obesity and Maternal Responsibility." *Critical Public Health* 21, no. 2 (2011): 179–90.

Meaney, Michael J. "Maternal Care, Gene Expression, and the Transmission of Individual Differences in Stress Reactivity across Generations." *Annual Review of Neuroscience* 24, no. 1 (2001): 1161–92.

Medzhitov, Ruslan. "Approaching the Asymptote: 20 Years Later." *Immunity* 30, no. 6 (2009): 766–75.

Medzhitov, Ruslan, Paula Preston-Hurlburt, and Charles A. Janeway. "A Human Homologue of the *Drosophila* Toll Protein Signals Activation of Adaptive Immunity." *Nature* 388, no. 6640 (1997): 394–97.

Mehler, Mark F. "Epigenetic Principles and Mechanisms Underlying Nervous System Functions in Health and Disease." *Progress in Neurobiology* 86, no. 4 (2008): 305–41.

Mekel-Bobrov, Nitzan, Sandra L. Gilbert, Patrick D. Evans, Eric J. Vallender, Jeffrey R. Anderson, Richard R. Hudson, Sarah A. Tishkoff, and Bruce T. Lahn. "Ongoing Adaptive Evolution of Aspm, a Brain Size Determinant in *Homo sapiens*." *Science* 309 (2005): 1720–22.

Meloni, Maurizio. "Biology without Biologism: Social Theory in a Postgenomic Age." *Sociology* (2013), Online First.

Michalak, P. "RNA World—the Dark Matter of Evolutionary Genomics." *Journal of Evolutionary Biology* 19, no. 6 (2006): 1768–74.

Miele, Frank. "En-twinned Lives." *Skeptic* 14 (2008): 27–37.

Milgram, Stanley. "The Small World Problem." *Psychology Today* 1 (1967): 60–67.

Miller, David. "Introducing the 'Gay Gene': Media and Scientific Representations." *Public Understanding of Science* 4 (1995): 269–84.

Milne, Richard. "Drawing Bright Lines: Food and the Futures of Biopharming." In *Nature after the Genome*, edited by Sarah Parry and John Dupré. Malden, MA: Wiley-Blackwell, 2010.

Milo, R., S. Shen-Orr, S. Itzkovitz, N. Kashtan, D. Chklovskii, and Uri Alon. "Network Motifs: Simple Building Blocks of Complex Networks." *Science* 298 (2002): 824–27.

Misteli, Tom. "Beyond the Sequence: Cellular Organization of Genome Function." *Cell* 128 (2007): 787–800.

The modENCODE Consortium, Sushmita Roy, Jason Ernst, Peter V. Kharchenko, Pouya Kheradpour, Nicolas Negre, Matthew L. Eaton, Jane M. Landolin, et al. "Identification of Functional Elements and Regulatory Circuits by *Drosophila* modENCODE." *Science* 330 (2010): 1787–97.

Mold, J. E., J. Michaëlsson, T. D. Burt, M. O. Muench, K. P. Beckerman, M. P. Busch, T.-H. Lee, D. F. Nixon, and J. M. McCune. "Maternal Alloantigens Promote the Development of Tolerogenic Fetal Regulatory T Cells In Utero." *Science* 322 (2008): 1562–65.

Monod, Jacques, and François Jacob. "Teleonomic Mechanisms in Cellular Metabolism, Growth, and Differentiation." *Cold Spring Harbor Symposia on Quantitative Biology* 26 (1961): 389–401.

Moore, David S. *The Dependent Gene: The Fallacy of "Nature vs. Nurture."* New York: Times Books, 2002.

Moore, Kelly. *Disrupting Science: Social Movements, American Scientists, and the Politics of the Military, 1945–1975.* Princeton, NJ: Princeton University Press, 2008.

Morange, Michel. "Post-genomics, between Reduction and Emergence." *Synthese* 151, no. 3 (2006): 355–60.

Morello-Frosch, Rachel, and Edmond D. Shenassa. "The Environmental 'Riskscape' and Social Inequality: Implications for Explaining Maternal and Child Health Disparities." *Environmental Health Perspectives* 114 (2006): 1150–53.

Morgan, Hugh D., Heidi G. E. Sutherland, David I. K. Martin, and Emma Whitelaw. "Epigenetic Inheritance at the Agouti Locus in the Mouse." *Nature Genetics* 23, no. 3 (1999): 314.

Müller-Wille, Staffan, and Hans-Jörg Rheinberger. *A Cultural History of Heredity*. Chicago: University of Chicago Press, 2012.

Murphy, Michelle. *Sick Building Syndrome and the Problem of Uncertainty: Environmental Politics, Technoscience, and Women Workers*. Durham, NC: Duke University Press, 2006.

Murray, Thomas. "Race, Ethnicity, and Science: The Haplotype Genome Project." *Hastings Center Report* 31, no. 5 (2001): 7.

Mustanski, B. S., M. G. Dupree, C. M. Nievergelt, S. Bocklandt, N. J. Schork, and D. H. Hamer. "A Genomewide Scan of Male Sexual Orientation." *Human Genetics* 116 (2005): 272–78.

National Academy of Sciences. *Exposure Science in the 21st Century: A Vision and a Strategy*. Washington, DC: National Academies Press, 2012.

National Human Genome Research Institute. "ENCODE Data Describes Function of Human Genome." September 5, 2012. http://www.genome.gov/27549810.

National Human Genome Research Institute. "National Human Genome Research Institute NIH Health Disparities Strategic Plan Fiscal Years 2004–2008." 2003. http://www.genome.gov/pages/research/der/derreportspublications/nhgri healthdisparitiesplan.pdf.

National Institutes of Health. "Common Fund Strategic Planning Social Media Summary." 2012. http://commonfund.nih.gov/planningactivities/socialmedia _summary.

National Institutes of Health. "Genes, Environment and Health Initiative Invests In Genetic Studies, Environmental Monitoring Technologies." *NIH News*, September 4, 2007. http://www.nih.gov/news/pr/sep2007/nhgri-04.htm.

National Institutes of Health. "Global Health Meetings" 2010. http://commonfund .nih.gov/globalhealth/researchmeeting.aspx.

National Institutes of Health. "NIH FY 2012 Online Performance Appendix." 2011. http://dpcpsi.nih.gov/sites/default/files/opep/document/FY_2012_NIH _Online_Performance_Appendix.pdf.

National Institutes of Health. "NIH Guide: Studies of the Ethical, Legal and Social Implications of Research into Human Genetic Variation." April 29, 1999. http://grants.nih.gov/grants/guide/rfa-files/RFA-HG-99-002.html.

National Institutes of Health. *The NIH Revitalization Act of 1993*. Vol. PL 103-43, 1993.

National Institutes of Health. "NIH Roadmap: Reengineering the Clinical Research Enterprise, Regional Translational Research Centers Interim Report." June 1, 2004. http://commonfund.nih.gov/sites/default/files/rtrc_interimreport.pdf.

National Institutes of Health. "Report of the First Community Consultation on the Responsible Collection and Use of Samples for Genetic Research." September 25–26, 2000. http://www.nigms.nih.gov/News/Reports/Archived Reports2003-2000/Pages/community_consultation.aspx.

National Institutes of Health. "RFA-CA-09-001: NIH-Supported Centers for Population Health and Health Disparities (CPHHD) (P50)." 2009. http://grants.nih .gov/grants/guide/rfa-files/RFA-CA-09-001.html.

National Institutes of Health. "Two NIH Initiatives Launch Intensive Efforts to Determine Genetic and Environmental Roots of Common Diseases." National Institute of Environmental Health Services, February 8, 2006, Press Release 06-03 edition.

National Institutes of Health. "Understanding Our Genetic Inheritance: The U.S. Human Genome Project. The First Five Years: Fiscal Years 1991–1995." April 1990. http://www.ornl.gov/sci/techresources/Human_Genome/project/5yrplan/summary.shtml.

National Institutes of Mental Health. *The Numbers Count: Mental Disorders in America*. Washington, DC: National Institutes of Mental Health, 2006.

National Research Council. *Ensuring the Integrity, Accessibility, and Stewardship of Research Data in the Digital Age*. Washington, DC: National Academies Press, 2009.

National Research Council. *Sharing Publication-Related Data and Materials: Responsibilities of Authorship in the Life Sciences*. Washington, DC: National Academies Press, 2003.

Nelkin, Dorothy, and M. Susan Lindee. *The DNA Mystique: The Gene as a Cultural Icon*. 1st ed. New York: Freeman, 1995.

Nelkin, Dorothy, and M. Susan Lindee. *The DNA Mystique: The Gene as a Cultural Icon*. 2nd ed. Ann Arbor: University of Michigan Press, 2000.

Nelson, Alondra. *Body and Soul: The Black Panther Party and the Fight against Medical Discrimination*. Minneapolis: University of Minnesota Press, 2011.

Nelson, Karen E., and Bryan A. White. "Metagenomics and Its Application to the Study of the Human Microbiome." In *Metagenomics: Theory, Methods, and Applications*, edited by Diana Marco, 171–82. Norwich: Horizon Scientific Press, 2010.

Nerlich, B., and I. Hellsten. "Genomics: Shifts in Metaphorical Landscape between 2000 and 2003." *New Genetics and Society* 23, no. 3 (2004): 255–68.

Newman, Mark. *Networks: An Introduction*. New York: Oxford University Press, 2010.

Newman, Mark. "The Structure and Function of Complex Networks." *SIAM Review* 45 (2003): 167–256.

News Editorial Staff. "The Runners-up." *Science* 302 (2003): 2039–45.

Newton-Cheh, C., and Joel N. Hirschhorn. "Genetic Association Studies of Complex Traits: Design and Analysis Issues." *Mutation Research* 573 (2005): 54–69.

Nielsen, Rasmus. "Genomics: In Search of Rare Human Variants." *Nature* 467, no. 7319 (2010): 1050–51.

Niewohner, Jorg. "Epigenetics: Embedded Bodies and the Molecularisation of Biography and Milieu." *BioSocieties* 6, no. 3 (2011): 279–98.

November, Joseph. *Biomedical Computing: Digitizing Life in the United States*. Baltimore: Johns Hopkins University Press, 2012.

Odum, Howard, and Eugene Odum. *Fundamentals of Ecology*. Minneapolis: University of Minnesota Press, 1953.

Ohno, Suzumu. "So Much "Junk" DNA in Our Genome." *Brookhaven Symposia in Biology* 23 (1972): 366–70.

Olden, Kenneth, Nicholas Freudenberg, Jennifer Dowd, and Alexandra E. Shields. "Discovering How Environmental Exposures Alter Genes Could Lead to New Treatments for Chronic Illnesses." *Health Affairs* 30 (2011): 833–41.

Olden, Kenneth, and Sandra L. White. "Health-Related Disparities: Influence of Environmental Factors." *Medical Clinics of North America* 89 (2005): 721–38.

O'Malley, Maureen A., and John Dupré. "Fundamental Issues in Systems Biology." *Bioessays* 27 (2005): 1270–76.

O'Malley, Maureen A., and John Dupré. "Size Doesn't Matter: Towards a More Inclusive Philosophy of Biology." *Biology and Philosophy* 22 (2007): 155–91.

O'Malley, Maureen A., Kevin C. Elliot, Chris Haufe, and Richard M. Burian. Philosophies of Funding." *Cell* 138 (2009): 611–15.

O'Malley, Maureen A., and Orkun Soyer. "The Roles of Integration in Molecular Systems Biology." *Studies in the History and the Philosophy of the Biological and Biomedical Sciences* 43 (2012): 58–68.

Orgel, Leslie E., and Francis H. Crick. "Selfish DNA: The Ultimate Parasite." *Nature* 284, no. 5757 (1980): 604–7.

Otis, Laura. *Networking: Communicating with Bodies and Machines in the Nineteenth Century.* Ann Arbor: University of Michigan Press, 2001.

Ottino, Julio M. "Engineering Complex Systems." *Nature* 427 (2004): 399.

Ozdemir, Vural, David S. Rosenblatt, Louise Warnich, Sanjeeva Srivastava, Ghazi O. Tadmouri, Ramy K. Aziz, Panga Jaipal Reddy, et al. 2011. "Towards an Ecology of Collective Innovation: Human Variome Project (HVP), Rare Disease Consortium for Autosomal Loci (RaDiCAL) and Data-Enabled Life Sciences Alliance (DELSA)." *Current Pharmacogenomics and Personalized Medicine* 9 (2011): 243–51.

Panofsky, Aaron. "Field Analysis and Interdisciplinary Science: Scientific Capital Exchange in Behavior Genetics." *Minerva* 49 (2011): 295–316.

Panofsky, Aaron. *Misbehaving Science: Controversy and the Development of Behavior Genetics.* Chicago: University of Chicago Press, 2014.

Panofsky, Aaron. "Rethinking Scientific Authority: Race Controversies in Behavior Genetics." In *Creating Authority*, edited by C. Calhoun and R. Sennett. New York: New York University Press, forthcoming.

Parke, Ross D. *A Century of Developmental Psychology.* Washington, DC: American Psychological Association, 1994.

Parker, John N., Nikki Vermeulen, and Bart Penders, eds. *Collaboration in the New Life Sciences.* Aldershot: Ashgate, 2010.

Parry, R. M., W. Jones, T. H. Stokes, J. H. Phan, R. A. Moffitt, H. Fang, L. Shi, et al. "K-Nearest Neighbor Models for Microarray Gene Expression Analysis and Clinical Outcome Prediction." *Pharmacogenomics Journal* 10, no. 4 (2010): 292–309.

Parry, Sarah. "Interspecies Entities and the Politics of Nature." In *Nature after the Genome*, edited by Sarah Parry and John Dupré. Malden, MA: Wiley-Blackwell, 2010.

Patel, Chirag J., Jayanta Bhattacharya, and Atul J. Butte. "An Environment-Wide Association Study (EWAS) on Type 2 Diabetes Mellitus." Edited by Baohong Zhang. *PLoS ONE* 5 (2010): e10746.

Patton, Paul. "Events, Becoming, and History." In *Deleuze and History*, edited by Jeffrey A. Bell and Claire Colebrook, 33–53. Edinburgh: Edinburgh University Press, 2009.

Paul, Annie Murphy. *Origins: How the Nine Months before Birth Shape the Rest of Our Lives*. New York: Free Press, 2010.

Pembrey, Marcus E., Lars Olov Bygren, Gunnar Kaati, Sören Edvinsson, Kate Northstone, Michael Sjöstrom, Jean Golding, and the ALSPAC Study Team. "Sex-Specific, Male-Line Transgenerational Responses in Humans." *European Journal of Human Genetics* 14, no. 2 (2005): 159–66.

Pennisi, Elizabeth. "Are Epigeneticists Ready for Big Science?" *Science* 319, no. 5867 (2008): 1177.

Pennisi, Elizabeth. "A Low Number Wins the GeneSweep Pool." *Science* 300, no. 5625 (2003): 1484–85.

Pennisi, Elizabeth. "Shining a Light on the Genome's 'Dark Matter.'" *Science* 330 (2010): 1614.

Pescosolido, Bernice A. "Of Pride and Prejudice: The Role of Sociology and Social Networks in Integrating the Health Sciences." *Journal of Health and Social Behavior* 47 (2006): 189–208.

Petronis, Arturas. "Epigenetics as a Unifying Principle in the Aetiology of Complex Traits and Diseases." *Nature* 465, no. 7299 (2010): 721–27.

Petryna, Adriana. *When Experiments Travel: Clinical Trials and the Global Search for Human Subjects*. Princeton, NJ: Princeton University Press, 2009.

Phillips, Kathryn A., David L. Veenstra, Eyal Oren, Jane K. Lee, and Wolfgang Sadee. "Potential Role of Pharmacogenomics in Reducing Adverse Drug Reactions: A Systematic Review." *Journal of the American Medical Association* 286 (2001): 2270–79.

Pickering, Andrew. *The Cybernetic Brain: Sketches of Another Future*. Chicago: University of Chicago Press, 2010.

Pickering, Andrew. "Cybernetics and the Mangle: Ashby, Beer, and Pask." *Social Studies of Science* 32 (2002): 413–37.

Pinker, Steven. "My Genome, My Self." *New York Times*, January 11, 2009.

Piwowar, Heather. "Who Shares? Who Doesn't? Factors Associated with Openly Archiving Raw Research Data." *PLoS One* 6 (2011): e18657.

Piwowar, Heather A., Roger S. Day, and Douglas B. Fridsma. "Sharing Detailed Research Data Is Associated with Increased Citation Rate." *PLoS One* 2, no. 3 (2007): e308.

Plant, Rebecca Jo. *Mom: The Transformation of Motherhood in Modern America*. Chicago: University of Chicago Press, 2010.

Plomin, Robert, John C. DeFries, Ian W. Craig, and Peter McGuffin. "Behavioral Genetics." In *Behavioral Genetics in the Postgenomic Era*, edited by R. Plomin,

J. C. DeFries, I. W. Craig, and P. McGuffin, 3–15. Washington, DC: American Psychological Association, 2003.

Plomin, Robert, John C. DeFries, Ian W. Craig, and Peter McGuffin. *Behavioral Genetics in the Postgenomic Era*. Washington, DC: American Psychological Association, 2003.

Plomin, Robert, Michael J. Owen, and Peter McGuffin. "The Genetic Basis of Complex Human Behaviors." *Science* 264 (1994): 1733–39.

Pohlhaus, Jennifer Reineke, and Robert M. Cook-Deegan. "Genomics Research: World Survey of Public Funding." *BMC Genomics* 9 (2008): 472.

Qureshi, Irfan A., and Mark F. Mehler. "Emerging Roles of Non-coding RNAs in Brain Evolution, Development, Plasticity and Disease." *Nature Reviews Neuroscience* 13, no. 8 (2012): 528–41.

Rabinow, Paul, and Nikolas Rose. "Biopower Today." *BioSocieties* 1 (2006): 195–217.

Radder, Hans, ed. *The Commodification of Academic Research: Science and the Modern University*. Pittsburgh: University of Pittsburgh Press, 2010.

Rapoport, Anatol. "Mathematical Models of Social Interaction." In *Handbook of Mathematical Psychology*, vol. 2, edited by R. D. Luce, R. R. Bush, and E. Galanter, 493–579. New York: Wiley, 1963.

Rapoport, Anatol. "Spread of Information through a Population with Sociostructural Bias: I. Assumption of Transitivity." *Bulletin of Mathematical Biophysics* 15 (1953): 523–33.

Rapp, Rayna. *Testing Women, Testing the Fetus: The Social Impact of Amniocentesis in America*. New York: Routledge, 2000.

Rappaport, Stephen M. "Biomarkers Intersect with the Exposome." *Biomarkers: Biochemical Indicators of Exposure, Response, and Susceptibility to Chemicals* 17 (2012): 483–89.

Rappaport, Stephen M. "Implications of the Exposome for Exposure Science." *Journal of Exposure Science and Environmental Epidemiology* 21 (2011): 5–9.

Rappaport, Stephen M., He Li, Hasmik Grigoryan, William E. Funk, and Evan R. Williams "Adductomics: Characterizing Exposures to Reactive Electrophiles." *Toxicology Letters* 213 (2012): 83–90.

Rappaport, Stephen M., and Martyn T. Smith. "Environment and Disease Risks." *Science* 330 (2010): 460–61.

Raser, J. M., and E. K. O'Shea. "Noise in Gene Expression: Origins, Consequences, and Control." *Science* 309 (2005): 2010–13.

Rashevsky, Nicolas. *Mathematical Theory of Human Relations: An Approach to the Mathematical Biology of Social Phenomena*. Bloomington, IN: Principia Press, 1947.

Rawls, J. F., B. S. Samuel, and J. I. Gordon. "Gnotobiotic Zebrafish Reveal Evolutionarily Conserved Responses to the Gut Microbiota." *PNAS* 101, no. 13 (2004): 4596–601.

RDevelopmentCoreTeam. "The R Project for Statistical Computing." 2010. http://www.r-project.org/.

Reardon, Jenny. "The Democratic, Anti-Racist Genome? Technoscience at the Limits of Liberalism." *Science as Culture* 21, no. 1 (2011): 25–47.

Reardon, Jenny. *Race to the Finish: Identity and Governance in an Age of Genomics*. Princeton, NJ: Princeton University Press, 2005.

Remenyi, Attila, Hans R. Scholer, and Matthias Wilmanns. "Combinatorial Control of Gene Expression." *Nature Structural and Molecular Biology* 11 (2004): 812–15.

Renear, Allan H., and Carole L. Palmer. "Strategic Reading, Ontologies, and the Future of Scientific Publishing." *Science* 325 (2009): 828–32.

Rheinberger, Hans-Jörg. "Experimental Systems, Graphematic Spaces." In *Inscribing Science: Scientific Texts and the Materiality of Communication*, edited by Timothy Lenoir, 285–303. Stanford, CA: Stanford University Press, 1998.

Rheinberger, Hans-Jörg. "Introduction to 'Writing Genomics.'" *History and Philosophy of the Life Sciences* 32, no. 1 (2010): 63–64.

Rheinberger, Hans-Jörg, and Jean-Paul Gaudilliere, eds. *Classical Genetic Research and Its Legacy: The Mapping Cultures of Twentieth Century Genetics*. New York: Routledge, 2004.

Rice, Clinton, Jugal K. Ghorai, Kathryn Zalewski, and Daniel N. Weber. "Developmental Lead Exposure Causes Startle Response Deficits in Zebrafish." *Aquatic Toxicology* 105, no. 3–4 (2011): 600–608.

Rice, G., C. Anderson, N. Risch, and G. Ebers. "Male Homosexuality: Absence of Linkage to Microsatellite Markers at Xq28." *Science* 284 (1999): 665–67.

Richardson, Sarah S. "Race and IQ in the Postgenomic Age: The Microcephaly Case." *BioSocieties* 6 (2011): 420–46.

Risch, Neil J., and David Botstein. "A Manic Depressive History." *Nature Genetics* 12 (1996): 351–53.

Risch, Neil, Esteban Burchard, Elad Ziv, and Hua Tang. "Categorization of Humans in Biomedical Research: Genes, Race and Disease." *Genome Biology* 3 (2002): 1–12.

Risch, Neil, Richard Herrell, Thomas Lehner, Kung-Yee Liang, Lindon Eaves, Josephine Hoh, Andrea Griem, Maria Kovacs, Jurg Ott, and Kathleen Ries Merikangas. "Interaction between the Serotonin Transporter Gene (5-Httlpr), Stressful Life Events, and Risk of Depression: A Meta-analysis." *Journal of the American Medical Association* 301 (2009): 2462–71.

Roberts, Eric M. "Neighborhood Social Environments and the Distribution of Low Birthweight in Chicago." *American Journal of Public Health* 87 (1997): 597–603.

Roberts, L. "How to Sample the World's Genetic Diversity." *Science* 257 (1992): 1204–5.

Robertson, Miranda. "False Start on Manic Depression." *Nature* 342 (1989): 222.

Rodriguez, A., J. Miettunen, T. B. Henriksen, J. Olsen, C. Obel, A. Taanila, H. Ebeling, K. M. Linnet, I. Moilanen, and M.-R. Järvelin. "Maternal Adiposity prior to Pregnancy Is Associated with ADHD Symptoms in Offspring." *International Journal of Obesity* 32, no. 3 (2008): 550–57.

Rogaev, Evgeny Ivanovich. "Genomics of Behavioral Diseases." *Frontiers in Genetics* 3 (2012): 1–4.

Rogers, Susan, and Alberto Cambrosio. 2007. "Making a New Technology Work: The Standardisation and Regulation of Microarrays." *Yale Journal of Biology and Medicine* 80:165–78.

Rose, Hilary, and Steven Rose. *Genes, Cells and Brains: The Promethean Promises of the New Biology.* London: Verso, 2013.

Rose, Nikolas S. "Normality and Pathology in a Biomedical Age." *Sociological Review* 57, no. s2 (2009): 66–83.

Rose, Nikolas S. *The Politics of Life Itself: Biomedicine, Power, and Subjectivity in the Twenty-First Century.* Princeton, NJ: Princeton University Press, 2007.

Rose, Steven P. R. "Neurogenetic Determinism and the New Euphenics." *British Medical Journal* 317 (1998): 1707–8.

Rose, Steven P. R. "The Rise of Neurogenetic Determinism." *Nature* 373 (1995): 380–82.

Roseboom, T. J. "Undernutrition during Fetal Life and the Risk of Cardiovascular Disease in Adulthood." *Future Cardiology* 8, no. 1 (2012): 5–7.

Rotimi, Charles N. "Health Disparities in the Genomic Era: The Case for Diversifying Ethnic Representation." *Genome Medicine* 4 (2012): 1–3.

Rotimi, Charles N., and L. B. Jorde. "Ancestry and Disease in the Age of Genomic Medicine." *New England Journal of Medicine* 363, no. 16 (2010): 1551–58.

Ruiz-Narváez, Edward A., and Lynn Rosenberg. "Validation of a Small Set of Ancestral Informative Markers for Control of Population Admixture in African Americans." *American Journal of Epidemiology* 173, no. 5 (2011): 587–92.

Rutter, Michael, and Robert Plomin. "Opportunities for Psychiatry from Genetic Findings." *British Journal of Psychiatry* 171 (1997): 209–19.

Sampson, Robert J. *The Great American City: Chicago and the Enduring Neighborhood Effect.* Chicago: University of Chicago Press, 2012.

Sampson, Robert J., and Jeffrey D. Morenoff. "Durable Inequality: Spatial Dynamics, Social Processes, and the Persistence of Poverty in Chicago Neighborhoods." In *Poverty Traps*, edited by Samuel Bowels, Steven Durlauf, and Karl Hoff. New York: Princeton University Press and Russell Sage Foundation, 2006.

Sampson, Robert J., Jeffrey D. Morenoff, and Thomas Gannon-Rowley. "Assessing 'Neighborhood Effects': Social Processes and New Directions in Research." *Annual Review of Sociology* 28 (2002): 443–78.

Sampson, Robert J., and William Julius Wilson. "Toward a Theory of Race, Crime, and Urban Inequality." In *Crime and Inequality*, edited by John Hagan and Ruth D. Peterson. Stanford, CA: Stanford University Press, 1995.

Sanders, I. R. "Ecology and Evolution of Multigenomic Arbuscular Mycorrhizal Fungi." *American Naturalist* 160 (2002): S128–S141.

Sankar, Pamela, Mildred K. Cho, Celeste M. Condit, Linda M. Hunt, Barbara Koenig, Patricia Marshall, Sandra Soo-Jin Lee, and Paul Spicer. "Genetic Research and Health Disparities." *Journal of the American Medical Association* 291 (2004): 2985–89.

Sapp, Jan. *Beyond the Gene: Cytoplasmic Inheritance and the Struggle for Authority in Genetics.* Oxford: Oxford University Press, 1987.

Schramek, Tania Elaine. "Michael Meaney: Ode to the Mother's Touch." *Douglas Mental Health University Institute News*, September 1, 2005. http://www.douglas.qc.ca/news/1006.

Schuster, Stephan C., Webb Miller, Aakrosh Ratan, Lynn P. Tomsho, Belinda Giardine, Lindsay R. Kasson, Robert S. Harris, et al. "Complete Khoisan and Bantu Genomes from Southern Africa." *Nature* 463, no. 7283 (2010): 943–47.

Schwartz, David, and Francis S. Collins. "Medicine: Environmental Biology and Human Disease." *Science* 316 (2007): 695–96.

Sebald, Hans. *Momism: The Silent Disease of America*. Chicago: Nelson-Hall, 1976.

Sedgwick, Eve Kosofsky, and Adam Frank. "Shame in the Cybernetic Fold: Reading Silvan Tomkins." In *Shame and Its Sisters: A Silvan Tomkins Reader*, edited by Eve K. Sedgwick and Adam Frank, 1–28. Durham, NC: Duke University Press, 1995.

Shanahan, Michael J., and Scott M. Hofer. "Social Context in Gene-Environment Interactions: Retrospect and Prospect." *Journals of Gerontology Series B: Psychological Sciences and Social Sciences* 60 (2005): 65–76.

Shapin, Steven. "The Invisible Technician." *American Scientist* 77 (1989): 554–64.

Shapin, Steven. *The Scientific Life: A Moral History of a Late Modern Vocation*. 1st ed. Chicago: University of Chicago Press, 2008.

Shaw, Clifford, and Henry McKay. *Juvenile Delinquency and Urban Areas*. Chicago: Chicago University Press, 1969.

Shiffrin, Richard M., and Katy Börner. "Mapping Knowledge Domains." *Proceedings of the National Academy of Sciences USA* 101 (2004): 5183–85.

Shorter, Edward. *A History of Psychiatry*. New York: John Wiley and Sons, 1997.

Shostak, Sara N. *Exposed Science: Genes, the Environment, and the Politics of Population Health*. Berkeley: University of California Press, 2013.

Shostak, Sara N., and Jeremy Freese. "Gene-Environment Interaction and Medical Sociology." In *Handbook of Medical Sociology*, edited by Chloe E. Bird, Peter Conrad, Allen M. Fremont, and Stefan Timmermans, 418–34. Nashville: Vanderbilt University Press, 2010.

Shubin, Neil. *Your Inner Fish: A Journey into the 3.5-Billion-Year History of the Human Body*. New York: Vintage, 2009.

Shulevitz, Judith. "Why Fathers Really Matter." *New York Times*, September 8, 2012.

Simon, Herbert. "A Formal Theory of Interaction in Social Groups." *American Sociological Review* 17 (1952): 202–12.

Siontis, Konstantinos C. M., Nikolaos A. Patsopoulos, and John P. A. Ioannidis. "Replication of Past Candidate Loci for Common Diseases and Phenotypes in 100 Genome-Wide Association Studies." *European Journal of Human Genetics* 18 (2010): 832–37.

Slikker, W., Jr. "Of Genomics and Bioinformatics." *Pharmacogenomics Journal* 10, no. 4 (2010): 245–46.

Smart, Andrew, Richard Tutton, Richard Ashcroft, Paul A. Martin, and George T. H. Ellison. "Can Science Alone Improve the Measurement and Communication of Race and Ethnicity in Genetic Research? Exploring the Strategies Proposed by Nature Genetics." *Biosocieties* 1 (2006): 313–24.

Smith, M. L., J. N. Bruhn, and J. B. Anderson. "The Fungus *Armillaria bulbosa* Is among the Largest and Oldest Living Organisms." *Nature* 356 (1992): 428–31.

Smith, Martyn T., Luoping Zhang, Cliona M. McHale, Christine F. Skibola, and Stephen M. Rappaport. "Benzene, the Exposome and Future Investigations of Leukemia Etiology." *Chemico-biological Interactions* 192 (2011): 155–59.

Smith, T. "The History of the Genetic Sequence Databases." *Genomics* 6 (1990): 701–7.

Somel, Mehmet, Xiling Liu, Lin Tang, Zheng Yan, Haiyang Hu, Song Guo, Xi Jiang, et al. "MicroRNA-Driven Developmental Remodeling in the Brain Distinguishes Humans from Other Primates." *PLoS Biology* 9, no. 12 (2011): e1001214.

Stankiewicz, Pawel, and James R. Lupski. "Structural Variation in the Human Genome and Its Role in Disease." *Annual Review of Medicine* 61 (2010): 437–55.

Star, Susan Leigh, and Karen Ruhleder. "Steps towards an Ecology of Infrastructure: Design and Access for Large Information Spaces." *Information Systems Research* 7 (1996): 111–34.

Stein, Lincoln D. "Towards a Cyberinfrastructure for the Biological Sciences: Progress, Visions and Challenges." *Nature Reviews Genetics* 9 (2008): 678–88.

Stengers, Isabelle. "Experimenting with *What Is Philosophy?*" In *Deleuzian Intersections: Science, Technology, Anthropology*, edited by Caspar Bruun Jensen and Kjetil Rodje. New York: Berghahn Books, 2010.

Sternberg, Richard. "On the Roles of Repetitive DNA Elements in the Context of a Unified Genomic-Epigenetic System." *Annals of the New York Academy of Sciences* 981 (2002): 154–88.

Stevens, Hallam. "Coding Sequences: A History of Sequence Comparison Algorithms as a Scientific Instrument." *Perspectives on Science* 19, no. 3 (2011): 263–99.

Stevens, Hallam. *Life out of Sequence*. Chicago: University of Chicago Press, 2013.

Stevens, Hallam. "On the Means of Bio-production: Bioinformatics and How to Make Knowledge in a High-Throughput Genomics Laboratory." *BioSocieties* 6, no. 2 (2011): 217–42.

Stoltenberg, Scott F., and Margit Burmeister. "Recent Progress in Psychiatric Genetics—Some Hope but No Hype." *Human Molecular Genetics* 9 (2000): 927–35.

Stoto, Michael A., Ruth Behrens, and Connie Rosemont. "Healthy People 2000: Citizens Chart the Course." 1990.

Stotz, Karola C. "The Ingredients for a Postgenomic Synthesis of Nature and Nurture." *Philosophical Psychology* 21, no. 3 (2008): 359–81.

Stotz, Karola C., A. Bostanci, and P. E. Griffiths. "Tracking the Shift to 'Postgenomics.'" *Community Genetics* 9, no. 3 (2006): 190–96.

Stranger, Barbara E., Eli A. Stahl, and Towfique Raj. "Progress and Promise of Genome-Wide Association Studies for Human Complex Trait Genetics." *Genetics* 187, no. 2 (2011): 367–83.

Strasser, Bruno J. "Collecting, Comparing, and Computing Sequences: The Making of Margaret O. Dayhoff's *Atlas of Protein Sequence and Structure*, 1954–1965." *Journal of the History of Biology* 43 (2010): 623–60.

Strasser, Bruno J. "Data-Driven Sciences: From Wonder Cabinets to Electronic Databases." *Studies in History and Philosophy of Biological and Biomedical Sciences* 43, no. 1 (2012): 85–87.

Strasser, Bruno J. "The Experimenter's Museum: Genbank, Natural History, and the Moral Economies of Biomedicine." *Isis* 102 (2011): 60–96.

Strasser, Bruno J. "GenBank: Natural History in the 21st Century?" *Science* 322 (2008): 537–38.

Strogatz, Steven. "Exploring Complex Networks." *Nature* 410 (2001): 268–76.

Sullivan, P. "Don't Give Up on GWAS." *Molecular Psychiatry* (2011).

Sullivan, P. F. "Spurious Genetic Associations." *Biological Psychiatry* 61 (2007): 1121–26.

"Summary of Principles Agreed upon at the First International Strategy Meeting on Human Genome Sequencing (Bermuda, 25–28 February 1996)." *Human Genome Project Information Archive 1990–2003.* http://web.ornl.gov/sci/tech resources/Human_Genome/research/bermuda.shtml#1.

Sunder Rajan, Kaushik. *Biocapital: The Constitution of Postgenomic Life.* Durham, NC: Duke University Press, 2006.

Susser, Mervyn. "The Logic in Ecological: I. The Logic of Analysis." *American Journal of Public Health* 84 (1994): 825–29.

Swarbreck, David, Christopher Wilks, Philippe Lamesch, Tanya Z. Berardini, Margarita Garcia-Hernandez, Hartmut Foerster, Donghui Li, et al. "The Arabidopsis Information Resource (TAIR): Gene Structure and Functional Annotation." *Nucleic Acid Research* 36 (2008): D1009–14.

Szallasi, Zoltan, Jorg Stelling, and Vipul Periwal. *System Modeling in Cellular Biology: From Concepts to Nuts and Bolts.* Cambridge, MA: MIT Press, 2006.

Szyf, M., I. C. Weaver, F. A. Champagne, J. Diorio, and M. J. Meaney. "Maternal Programming of Steroid Receptor Expression and Phenotype through DNA Methylation in the Rat." *Frontiers in Neuroendocrinology* 26, no. 3–4 (2005): 139–62.

Taft, Ryan J., Ken C. Pang, Timothy R. Mercer, Marcel Dinger, and John S. Mattick. "Non-coding RNAs: Regulators of Disease." *Journal of Pathology* 220, no. 2 (2010): 126–39.

Taylor, Chris F., D. Field, S. Sansone, J. Aerts, R. Apweiler, R., and Michael Ashburner. 2008. "Promoting Coherent Minimum Reporting Guidelines for Biological and Biomedical Investigations: The MIBBI Project." *Nature Biotechnology* 26:889–96.

Taylor, Peter. "Three Puzzles and Eight Gaps: What Heritability Studies and Critical Commentaries Have Not Paid Enough Attention To." *Biology and Philosophy* 25 (2010): 1–31.

Tenenbaum, Jessica D., Susanna-Assunta Sansone, and Melissa Haendel. "A Sea of Standards for Omics Data: Sink or Swim?" *Journal of the American Medical Informatics Association* (2013), Online First. doi:10.1136/amiajnl-2013–002066.

Thayer, Z. M., and C. W. Kuzawa. "Biological Memories of Past Environments: Epigenetic Pathways to Health Disparities." *Epigenetics* 6, no. 7 (2011): 798–803.

Thomas, C. A., Jr. "The Genetic Organization of Chromosomes." *Annual Review of Genetics* 5, no. 1 (1971): 237–56.

Thompson, Claire, Steven Cummins, Tim Brown, and Rosemary Kyle. "Understanding Interactions with the Food Environment: An Exploration of Supermarket Food Shopping Routines in Deprived Neighbourhoods." *Health and Place* 19 (2013): 116–23.

Tomkins, Silvan S. *Affect, Imagery, Consciousness: The Complete Edition.* New York: Springer, 2008.

Tsotsis, Alexia. "Another $50M Richer, 23andMe Drops Its Price To $99 Permanently." *TechCrunch*, December 11, 2012. http://techcrunch.com/2012/12/11/23andnotme/.

Tuch, Brian B., David J. Galgoczy, Aaron D. Hernday, Hao Li, and Alexander D. Johnson. "The Evolution of Combinatorial Gene Regulation in Fungi." *PLoS Biology* 6 (2008): e38.

Turkheimer, Eric. "Commentary: Variation and Causation in the Environment and Genome." *International Journal of Epidemiology* 40 (2011): 598–601.

Turkheimer, Eric. "Genome Wide Association Studies of Behavior Are Social Science." In *Philosophy of Behavioral Biology*, edited by Kathryn S. Plaisance and Thomas A. C. Reydon, 43–64. New York: Springer, 2012.

Turkheimer, Eric. "Mobiles: A Gloomy View of the Future of Research into Complex Human Traits." In *Wrestling with Behavioral Genetics*, edited by E. Parens, A. Chapman, and N. Press, 165–78. Baltimore: Johns Hopkins University Press, 2006.

Turkheimer, Eric. "Spinach and Ice Cream: Why Social Science Is So Difficult." In *Behavior Genetics Principles: Perspectives in Developments, Personality, and Psychopathology*, edited by L. DiLalla, 161–89. Washington, DC: American Psychological Association, 2004.

Turkheimer, Eric, Andreana Haley, Mary Waldron, Brian D'Onofrio, and Irving I. Gottesman. "Socioeconomic Status Modifies Heritability of IQ in Young Children." *Psychological Science* 14 (2003): 623–25.

Tutton, Richard. "Biobanks and the Inclusion of Racial/Ethnic Minorities." *Race/Ethnicity: Multidisciplinary Global Perspectives* 3 (2009): 75–95.

Tuv, E., A. Borisov, G. Runger, and K. Torkkola. "Feature Selection with Ensembles, Artificial Variables, and Redundancy Elimination." *Journal of Machine Learning Research* 10 (2009): 1341–66.

Twine, Richard. "Genomic Natures Read through Posthumanisms." In *Nature after the Genome*, edited by Sarah Parry and John Dupré. Malden, MA: Wiley-Blackwell, 2010.

Uetz, Peter, Loic Giot, Gerard Cagney, Traci A. Mansfield, Richard S. Judson, James R. Knight, Daniel Lockshon, et al. "A Comprehensive Analysis of Protein-Protein Interactions in Saccharomyces Cerevisiae." *Nature* 403 (2000): 623–27.

Uexküll, Jakob von. *Theoretical Biology.* New York: Harcourt Brace, 1926.

UNESCO. "Bioethics and Human Population Genetics Research." 1995.

UNESCO. "Human Genome Diversity Project, Address Delivered by Luca Cavalli-Sforza, Stanford University, to a Special Meeting of UNESCO." September 12, 1994. http://www.osti.gov/scitech/servlets/purl/505327.

U.S. Department of Energy (DOE) and the Human Genome Project. "To Know Our-selves: The U.S. Department of Energy and the Human Genome Project." July 1996. http://www.ornl.gov/sci/techresources/Human_Genome/publicat/tko/.

U.S. Department of Health and Human Services. "FY 2012 President's Budget for HHS." 2011. http://www.hhs.gov/about/hhsbudget.html.

U.S. Department of Health and Human Services. "FY 2013 President's Budget for HHS." 2012. http://www.hhs.gov/budget/fy2013/index.html.

U.S. Department of Health and Human Services. "HHS Strategic Plan, 2007–2012." 2007. http://aspe.hhs.gov/hhsplan/2007/.

U.S. Department of Health and Human Services. "Strategic Research Plan and Budget to Reduce and Ultimately Eliminate Health Disparities Volume I: Fiscal Years 2002–2006." 2000. http://www.nimhd.nih.gov/our_programs/strategic/pubs/VolumeI_031003EDrev.pdf.

U.S. Department of Health and Human Services. "Tracking Healthy People 2010." November 2000. http://www.healthypeople.gov/2010/document/tableofcontents.htm#tracking.

U.S. Office of Management and Budget. "OMB Directive No. 15." OMB Publications Office, 1978.

van den Oord, Edwin J. C. G., and Patrick F. Sullivan. "False Discoveries and Models for Gene Discovery." *Trends in Genetics* 19 (2003): 537–42.

Venter, J. Craig. "Multiple Personal Genomes Await." *Nature* 464, no. 7289 (2010): 676–77.

Visscher, Peter M., Matthew A. Brown, Mark I. McCarthy, and Jian Yang. "Five Years of GWAS Discovery." *American Journal of Human Genetics* 90, no. 1 (2012): 7–24.

Vogelstein, Bert, David Lane, and Arnold J. Levine. "Surfing the p53 Network." *Nature* 408 (2000): 307–10.

von Eschenbach, Andrew C., and Kenneth H. Buetow. "Cancer Informatics Vision: caBIG." *Cancer Informatics* 2 (2006): 22–24.

Wade, Nicholas. "A Decade Later, Genetic Map Yields Few New Cures." *New York Times,* June 12, 2010.

Wade, Nicholas. "Genome Mappers Navigate the Tricky Terrain of Race." *New York Times,* July 20, 2001. http://www.nytimes.com/2001/07/20/science/20GENO.html.

Wade, Nicholas. "Race Is Seen as Real Guide to Track Roots of Disease." *New York Times,* July 30, 2002. http://www.nytimes.com/2002/07/30/science/race-is-seen-as-real-guide-to-track-roots-of-disease.html.

Wade, Nicholas. "Toward the First Racial Medicine." *New York Times,* November 13, 2004. http://www.nytimes.com/2004/11/13/opinion/13sat2.html.

Wadhwa, Pathik D., Claudia Buss, Sonja Entringer, and James M. Swanson. "Developmental Origins of Health and Disease: Brief History of the Approach and Current Focus on Epigenetic Mechanisms." *Seminars in Reproductive Medicine* 27, no. 5 (2009): 358–68.

Warin, Megan, Tanya Zivkovic, Vivienne Moore, and Michael Davies. "Mothers as Smoking Guns: Fetal Overnutrition and the Reproduction of Obesity." *Feminism and Psychology* 22, no. 3 (2012): 360–75.

Warner, Judith. *Perfect Madness: Motherhood in the Age of Anxiety*. New York: Riverhead Books, 2005.

Watson, James D. Address at Stated Meeting of the American Academy of Arts and Sciences, Cambridge, Massachusetts, February 14, 1990. Author's transcript.

Watson, James D. *Avoid Boring People: Lessons from a Life in Science*. New York: Knopf, 2007.

Watson, James D. "A Molecular Genetics Perspective." In *Behavioral Genetics in the Postgenomic Era*, edited by R. Plomin, J. C. DeFries, I. W. Craig, and P. McGuffin, xxi–xxii. Washington, DC: American Psychological Association, 2003.

Watson, James D. *Recombinant DNA: Genes and Genomes: A Short Course*. Vol. 3. New York: W. H. Freeman, 2007.

Watts, Duncan J. "The 'New' Science of Networks." *Annual Reviews of Sociology* 30 (2004): 243–70.

Watts, Duncan J. *Six Degrees: The Science of a Connected Age*. New York: Norton, 2003.

Watts, Duncan J., and Steven H. Strogatz. "Collective Dynamics of 'Small World' Networks." *Nature* 393 (1998): 440–42.

Weaver, Ian C. G., Michael J. Meaney, and Moshe Szyf. "Maternal Care Effects on the Hippocampal Transcriptome and Anxiety-Mediated Behaviors in the Offspring That Are Reversible in Adulthood." *Proceedings of the National Academy of Sciences* 103, no. 9 (2006): 3480–85.

Weinberg, Robert A. "There Are Two Large Questions." *Debate* 5 (1991): 78.

Weiss, Rick. "Genome Project Completed; Findings May Alter Humanity's Sense of Itself, Experts Predict." *Washington Post*, April 15, 2003.

Weissmann, Gerald. "Pattern Recognition and Gestalt Psychology: The Day Nusslein-Volhard Shouted 'Toll!'" *FASEB Journal* 24 (2010): 2137–41.

Wells, Jonathan C. K. "A Critical Appraisal of the Predictive Adaptive Response Hypothesis." *International Journal of Epidemiology* 41, no. 1 (2012): 229–35.

Wells, Jonathan C. K. "Maternal Capital and the Metabolic Ghetto: An Evolutionary Perspective on the Transgenerational Basis of Health Inequalities." *American Journal of Human Biology* 22, no. 1 (2010): 1–17.

Wells, Jonathan C. K. "The Thrifty Phenotype as an Adaptive Maternal Effect." *Biological Reviews* 82, no. 1 (2007): 143–72.

Westerhoff, Hans V., and Bernhard O. Palsson. "The Evolution of Molecular Biology into Systems Biology." *Nature Biotechnology* 22 (2004): 1249–52.

Whitlock, Michael C., Mark A. McPeek, Mark D. Rausher, Loren Rieseberg, and Allen J. Moore. "Data Archiving." *American Naturalist* 175 (2010): 145–46.

Whitmarsh, Ian. *Biomedical Ambiguity: Race, Asthma, and the Contested Meaning of Genetic Research in the Caribbean*. Ithaca, NY: Cornell University Press, 2008.

Wiener, Norbert. *Cybernetics, or Communication and Control in the Animal and Machine*. Cambridge, MA: MIT Press, 1948.

Wild, Christopher P. "Complementing the Genome with an 'Exposome': The Outstanding Challenge of Environmental Exposure Measurement in Molecular Epidemiology." *Cancer Epidemiology Biomarkers and Prevention* 14 (2005): 1847–50.

Wild, Christopher P. "The Exposome: From Concept to Utility." *International Journal of Epidemiology* 41 (2012): 24–32.

Williams, George C. *Adaptation and Natural Selection.* Princeton, NJ: Princeton University Press, 1966.

Wilson, Elizabeth. "Scientific Interest: Introduction to Isabelle Stengers, 'Another Look: Relearning to Laugh.'" *Hypatia: A Journal of Feminist Philosophy* 15, no. 4 (2000): 38–40.

Wilson, Robert A. *Species: New Interdisciplinary Essays.* Cambridge, MA: MIT Press, 1999.

Wise, M. Norton. "Making Visible." *Isis* 97 (2006): 75–82.

Wojcicki, Anne. "23andMe Provides an Update regarding FDA's Review." *23andMe Blog,* December 5, 2013. http://blog.23andme.com/news/23andme-provides-an-update-regarding-fdas-review/.

Wolkenhauer, Olaf. "Systems Biology: The Reincarnation of Systems Theory Applied in Biology?" *Briefings in Bioinformatics* 2 (2001): 258–70.

Wong, Chloe Chung Yi, Avshalom Caspi, Benjamin Williams, Ian W. Craig, Renate Houts, Antony Ambler, Terrie E. Moffitt, and Jonathan Mill. "A Longitudinal Study of Epigenetic Variation in Twins." *Epigenetics* 5 (2010): 516–26.

Woolf, S. H. "The Meaning of Translational Research and Why It Matters." *Journal of the American Medical Association* 299, no. 2 (2008): 211–13.

Wouters, Paul, and Peter Schröeder. *The Public Domain of Digital Research Data.* Amsterdam: NIKI-KNAW, 2003.

Wright, R. J. "Prenatal Maternal Stress and Early Caregiving Experiences: Implications for Childhood Asthma Risk." *Paediatric and Perinatal Epidemiology* 21 (2007): 8–14.

Wu, T. T., Y. F. Chen, T. Hastie, E. Sobel, and K. Lange. "Genome-Wide Association Analysis by Lasso Penalized Logistic Regression." *Bioinformatics* 25, no. 6 (2009): 714–21.

Wynne, Brian. "Reflexing Complexity: Post-genomic Knowledge and Reductionist Returns in Public Science." *Theory, Culture and Society* 22, no. 5 (2005): 67–94.

Xie, H. G., R. B. Kim, A. J. Wood, and C. M. Stein. "Molecular Basis of Ethnic Differences in Drug Disposition and Response." *Annual Review of Pharmacology and Toxicology* 41 (2001): 815–50.

Yen, Irene H., and George A. Kaplan. "Neighborhood Social Environment and Risk of Death: Multilevel Evidence from the Alameda County Study." *American Journal of Epidemiology* 149 (1999): 898–907.

Yu, N., M. S. Kruskall, J. J. Yunis, H. M. Knoll, L. Uhl, S. Alosco, M. Ohashi, et al. "Disputed Maternity Leading to Identification of Tetragametic Chimerism." *New England Journal of Medicine* 346 (2002): 1545–52.

Zammit, Stanley N., Michael J. Owen, and Glyn Lewis. "Misconceptions about Gene-Environment Interactions in Psychiatry." *Evidence Based Mental Health* 13 (2010): 65–68.

Zammit, Stanley, N. Wiles, and Glyn Lewis. "The Study of Gene-Environment Interactions in Psychiatry: Limited Gains at Substantial Cost?" *Psychological Medicine* 40 (2010).

Zhang, T., L. C. Buoen, B. E. Seguin, G. R. Ruth, and A. F. Weber. "Diagnosis of Freemartinism in Cattle: The Need for Clinical and Cytogenic Evaluation." *Journal of the American Veterinary Medicine Association* 204, no. 10 (1994): 1672–75.

Zhang, Tie-Yuan, and Michael J. Meaney. "Epigenetics and the Environmental Regulation of the Genome and Its Function." *Annual Review of Psychology* 61, no. 1 (2010): 439–66.

Zimmer, Carl. "Now: The Rest of the Genome." *New York Times*, November 11, 2008.

Zimmer, Carl. "Yet-Another-Genome Syndrome." *Discover Magazine Blogs*, April 2, 2010. http://blogs.discovermagazine.com/loom/2010/04/02/yet-another -genome-syndrome/.

CONTRIBUTORS

RUSS ALTMAN

http://helix-web.stanford.edu/people/altman/

Russ Altman, MD, PhD, is a scientist at Stanford University Medical School, where he is a professor of bioengineering, genetics, and medicine, and of computer science by courtesy. He is chair of the Department of Bioengineering and director of the program in Biomedical Informatics. Altman's research focuses on the application of bioinformatics to basic molecular biological problems. Since the inception of the Human Genome Project, Altman has played a leading role in the development of genomics database and bioinformatics technologies and of the field of pharmacogenomics. He is a past president and one of the founding members of the International Society for Computational Biology. He is the principal investigator for the Pharmacogenomics Knowledgebase, a database that curates knowledge about the impact of genetic variation on drug response for clinicians and researchers. He is also the principal investigator for the Iranian Genome Project. Altman received his BA in Biochemistry and Molecular Biology from Harvard College and his MD and PhD in Medical Information Sciences from Stanford.

RACHEL ANKENY

http://www.adelaide.edu.au/directory/rachel.ankeny

Rachel A. Ankeny is Professor in the School of Humanities at the University of Adelaide. She is an interdisciplinary teacher and scholar whose areas of expertise cross three fields: history/philosophy of science, bioethics and science policy, and food studies. Prior to joining the University of Adelaide in 2006, she was director and lecturer/senior lecturer in the Unit for History and Philosophy of Science at the University of Sydney from 2000. In the history and philosophy of science, her research focuses on the roles of models and case-based reasoning in science, model organisms, the philosophy of medicine, and the history of contemporary life sciences. She also has expertise and ongoing research on health and science policy, particularly regarding public engagement.

CATHERINE BLISS

http://www.catherinebliss.com/

Catherine Bliss is Assistant Professor of Sociology at the University of California San Francisco. Her research explores the sociology of race, gender, and sexuality in medicine, and scientific controversies in genetics. Bliss's award-winning book *Race Decoded: The Genomic Fight for Social Justice* (2012) examines how genomics became today's new science of race. Her latest book project examines convergences in social and genetic science in the postgenomic age.

JOHN DUPRÉ

http://socialsciences.exeter.ac.uk/sociology/staff/dupre/

John Dupré is Professor of Philosophy of Science at the University of Exeter and the director of Egenis, the Centre for the Study of Life Sciences. He is the coauthor, most recently, of *Genomes and What to Make of Them*. He is a fellow of the American Association for the Advancement of Science, former President of the British Society for the Philosophy of Science, and a former member of the Governing Board of the Philosophy of Science Association.

MIKE FORTUN

http://www.sts.rpi.edu/pl/faculty/michael-fortun

Mike Fortun is an associate professor in the Department of Science and Technology Studies at Rensselaer Polytechnic Institute, Troy, New York. He was coeditor (with Kim Fortun) of *Cultural Anthropology*, the journal of the Society for Cultural Anthropology of the American Anthropological Association, from 2007 to 2010. A historian of the life sciences, his current research focuses on the contemporary science, culture, and political economy of genomics. His work in the life sciences has covered the policy, scientific, and social history of the Human Genome Project in the United States, the history of biotechnology, and the growth of commercial genomics and bioinformatics in the speculative economies of the 1990s. His most recent work is *Promising Genomics: Iceland and deCODE Genetics in a World of Speculation* (2008), an ethnographic account of deCODE Genetics in Iceland. He is currently part of an effort to build a digital humanities platform that will support collaborative ethnographic research into the science and care of asthma.

EVELYN FOX KELLER

http://web.mit.edu/sts/people/keller.html

Evelyn Fox Keller is Professor of the History and Philosophy of Science, Emerita, at the Massachusetts Institute of Technology. Her research focuses on the history and philosophy of modern biology and on gender and science. She is the author of several books, including *A Feeling for the Organism: The Life and Work of Barbara McClintock* (1983), *Reflections on Gender and Science* (1985), *The Century of the Gene* (2000), and *Making Sense of Life: Explaining Biological Development with Models, Metaphors and Machines* (2002). Her most recent book is *The Mirage of a Space between Nature and Nurture* (2010). Keller has received numerous academic and professional honors, including the Blaise Pascal Research Chair by the Préfecture de la Région D'Ile-de-France and elected membership in the American Philosophical Society and the American Academy of Arts and Science.

SABINA LEONELLI

http://socialsciences.exeter.ac.uk/sociology/staff/leonelli

Sabina Leonelli is Associate Professor in the Department of Sociology, Philosophy and Anthropology at the University of Exeter, U.K., where she also acts as Associate Director of the Exeter Centre for the Study of the Life

Sciences (Egenis). As an empirical philosopher of science, she uses historical and sociological research to foster philosophical understandings of knowledge-making practices and processes, focusing particularly on data-intensive biomedicine, model organism biology, and plant science. From 2014 to 2019, she holds an ERC Starting Grant to pursue research in this area (http://www.datastudies.eu). She is also interested in science policy, particularly current debates on Open Science. She has authored numerous papers in philosophy, sociology, and science journals; edited the volume *Scientific Understanding: Philosophical Perspectives* (2009) as well as special issues in *BioSocieties, Studies in the History and the Philosophy of the Biological and Biomedical Sciences*, and *Public Culture*. She is a member of several science and philosophy organizations, most notably the Global Young Academy.

ADRIAN MACKENZIE

http://www.lancaster.ac.uk/sociology/profiles/adrian-mackenzie

Adrian Mackenzie is Professor in Technological Cultures in the Department of Sociology and codirector of the Centre for Science Studies at Lancaster University. He is the author of *Transductions: Bodies and Machines at Speed*, (2002); *Cutting Code: Software and Sociality* (2006), and *Wirelessness: Radical Empiricism in Network Cultures* (2010). He is currently working on the circulation of data intensive methods across science, government, and business in network media.

MARGOT MOINESTER

http://sociology.fas.harvard.edu/people/margot-moinester

Margot Moinester is a PhD student in Sociology and a Doctoral Fellow in the Multidisciplinary Program in Inequality and Social Policy at Harvard University. Her research interests include health inequalities, immigration policy and enforcement, and crime and punishment.

AARON PANOFSKY

http://socgen.ucla.edu/people/aaron-panofsky/

Aaron Panofsky is Associate Professor in UCLA's Department of Public Policy and Institute for Society and Genetics. A sociologist of science, he recently published *Misbehaving Science: Controversy and the Development of Behavior Genetics* (2014). His other research interests include controversies over the genetics of race and the ways patient advocacy and internet technologies are transforming public participation in science.

SARAH S. RICHARDSON

http://scholar.harvard.edu/srichard

Sarah S. Richardson is John L. Loeb Associate Professor of the Social Sciences at Harvard University, jointly appointed in the History of Science and Studies of Women, Gender, and Sexuality. A historian and philosopher of science, her research focuses on race and gender in the biosciences and on the social dimensions of scientific knowledge. Richardson is the author of *Sex Itself: The Search for Male and Female in the Human Genome* (2013) and coeditor of *Revisiting Race in a Genomic Age* (2008).

SARA SHOSTAK

http://www.brandeis.edu/departments/sociology/people/faculty /shostak.html

Sara Shostak is Associate Professor of Sociology and Chair of the Health: Science, Society, and Policy Program at Brandeis University. Her research and teaching interests include the sociology of health and illness, science and technology studies, and environmental sociology. Across these domains, she focuses on how to understand and address inequalities in health. Her book *Exposed Science: Genes, the Environment, and the Politics of Population Health* (2013) received the Robert K. Merton Book Award and the Eliot Freidson Outstanding Publication Award.

HALLAM STEVENS

http://research.ntu.edu.sg/expertise/academicprofile/pages /StaffProfile.aspx?ST_EMAILID=HSTEVENS

Hallam Stevens is an Assistant Professor of History at Nanyang Technological University in Singapore. His research focuses on the impact of computers and information and communications technologies on science. His book *Life out of Sequence: Bioinformatics and the Introduction of Computers into Biology* (2013) is a historical and ethnographic account of the changes wrought to biological practice and biological knowledge by the introduction of the computer.

Caspi, Avshalom, 158, 166–67
Centers for Disease Control and Prevention (CDC), 177–78, 182
Center for Research on Genomics and Global Health, 182
Centers for Population Health and Health Disparities, 181–82
Centers of Excellence in Ethical Research, 182
Charney, Evan, 167
Chen, Xi, 82
chimerism, 62–65, 69
Christakis, Nicholas, 199
Circos representation, 73, 115, *116*
Clarke, Adele, 214
clones, 60–62
Cole, Steve, 24–25
Collins, Francis, 6, 9, 11, 181, 183, 192
complexity talk, 162–63
Conrad, Peter, 158
Coyne, Jerry, 4
credit attribution, 137–40, 145–47
Crick, Francis, 4, 15, 51, 126
cybernetics, 109

Daniels, Cynthia, 222
data, genomic, 73–75; access to, 128–29, 143; authorship and, 127, 138–39, 140, 144; handling of, 36; reuse of, 126–27, 136; RF-ACE algorithm, 75–77, 80–81, 82; segmentation, 96; from in situ hybridization, 134, *135*; SOM, 96–98; standards in, 132–33; statistical methods, 82–88; transformations in, 98–101; use tracking, 145–46; variations, 91–94
databases, 3, 34; bioinformatic, 78–81, 98–99, 237; data curation, 130, 132, 133–37, 140, 141–42, 146–47; data donation, 130–33, 143–44; metadata, 136–37, 144–45, 146; public/online, 84; 126–27, 128–29, 130, 131, 139–40. *See also* authorship; Bermuda Principles; credit attribu-

tion; maching learning techniques; specific databases
Davis, Bernard, 15–16, 45
Dawkins, Richard, 15, 104
Dayhoff, Margaret, 130
Department of Energy, U.S., genomics programs, 43, 44–45, 174, 177, 178, 182
Department of Health, U.K., 176
Department of Health and Human Services, U.S., 178, 180, 183, 188
depression, research on, 158, 166–67
developmental origins of health and disease (DOHaD), 217–18; epigenetic vectors in, 221–25; researcher profiles, 218–21
Dick, Danielle, 166
diet, study of, 203
digital object identifier (DOI), 144–45
DNA sequences, 3, 237; behavior and, 152; as boring, 39–40, 45; data sets, 81, 131; environmental signals and, 24, 27–28; epigenetic modification, 66; as genome, 18, 58–59, 65–66, 99–100; genomic disorders and, 22–23; haplotype map of, 21; junk, 5, 10, 11, 15–16; metaphors of, 104–5; methylation of, 57, 66–67, 212; mutations, 21–23, 27–28, 57; ratio of noncoding to total, *19*; regulation in, 14–15; selfish, 15
DOE. *See* Department of Energy, U.S.
DOHaD. *See* developmental origins of health and disease
DOI. *See* digital object identifier
Domenici, Pete, 45
Drosophila genetics, 13, 32–34
Dryad Digital Repository, 144–45

EBI. *See* European Bioinformatics Institute
Egeland, Janice, 157
ELSI. *See* Ethical, Legal, and Social Issues (ELSI) program

genomes: adaptability of, 23–24, 25, 219–20, 224; Circos representation of, 73, 115, *116*; complexity of, 6, 117–18, 119–20, 169, 234; dark matter of, 12, 17; data sets, 36, 73–74, 75–81; defined, 15, 16–17, 26–27, 56; diversity in, 57–58, 59–60; as essence of organism, 58–60, 65–66, 70; as foundational, 233; networked, 115; phylogenies of, 57, 67; plasticity of, 11, 215–16, 220; reactive, 10–11, 29; relation to genes, 9–10, 11–13; sequencing services, 233–34; structural changes of, 23; technologies, 3; transdisciplinary approach to, 237; unknown unknowns in, 2; use of term, 10, 16. *See also* postgenomics
genome-wide association studies (GWAS), 1–2, 3, 151, 152, 159; on chronic diseases, 196; on common variants, 241n18; microarrays, 83–86; predictors in, 87, 92, 99; statistical practices, 83–88; weak results in, 161–62, 193
GEO. *See* Gene Expression Omnibus
Gershon, Eliot, 159
Gibbs, W. Wayt, 12
Gilbert, Walter, 3, 39–41, 154
Global Health Initiative, 183
Gluckman, Peter, 222
GMOs. *See* genetically modified organisms
Google Compute Engine, 73, 74, 75, 100
Gottesman, Irving, 161
Gould, Todd, 161
Greally, John, 18
GWAS. *See* genome-wide association studies

Haeckel, Ernst, 116
Hamer, Dean, 154, 157–58
Hanson, Mark, 222
haplotype map (HapMap), 21, 180, 238
Harpending, Henry, 169

Hastie, Trevor, 94, 96
health disparities, 174–75, 177–88, 194–96, 200–205, 238
Health Disparities Plan (NHGRI), 182
health justice: defined in genetics, 179–81; environment and, 194, 204; funding for, 174, 180, 181–82, 188; introduction of concept, 174–75, 177; leadership in, 184, 187–88; in postgenomics, 177–86, 187; priorities, 181–84
Healthy People plan, 177
height, heritable trait, 2
Helmreich, Stefan, 169–70
heritability: behavior genetics on, 150–51, 157–59; DNA mutations and, 27–28; of IQ, 28, 159, 166; missing, 2, 162
HGP. *See* Human Genome Project
HHS. *See* Department of Health and Human Services, U.S.
high-throughput data collection, 3, 6, 74, 81, 120–21, 127, 180, 237, 240
Hilgartner, Stephen, 130, 132, 142
Hirschhorn, Joel, 158–59
Hoffman, Jules, 33
Holden, Constance, 158
holism, 3, 111, 121–22, 163–68, 214–15, 227, 235, 239
Hölzle, Urs, 73, 75
Hood, Leroy, 91
Human Epigenome Project, U.S., 213
Human Genome Diversity Project, 178
Human Genome Project (HGP), 1, 7, 11, 16, 33–34, 37, 56, 111, 192–93, 232, 239; behavior genetics and, 153–54; data sharing in, 128, 130; health justice in, 174, 177, 178, 179; maps in, 96; surprises of, 46–49
Human Heredity and Health in African Project, 183, 187

Ideker, Trey, 118
Illumina DNA sequencing machines, 233

Pharmacogenomics Research Network, 180–81

phenotypes, 15–16, 56–57, 91, 117, 220, 222, 233

Plomin, Robert, 160

Polymorphism Discovery Project, 178–79

postgenomics: in behavior genetics, 168–70; boring projects, 39–45; environment and, 25–29, 192–95, 202–5; epigenetics and, 213–17; evolution and, 23–25, 169–70; excitement in, 36–39, 51–53; goals of, 233; governance and, 175–77; health justice in, 177–86, 188; history of, 35–36; as holistic, 3, 122, 163–68, 214, 235; as interdisciplinary, 122; networks in, 121; state of science in, 186–87; style of thought, 184–86; surprises of, 46–49; tensions in, 235; transdisciplinary approach to, 236–38; use of term, 2–3, 5–6, 127–28, 232

prokaryotes, 17–18

Protein Data Bank, 129

public health, 174–75, 176, 177–80, 182, 187–88, 192, 210–11, 213, 218–19, 221–22

Public Health Department, U.S., 175

QTLs. *See* quantitative trait loci

quantitative trait loci (QTLs), 151

Qureshi, Irfan, 23–24

race: health inequality and, 174–75, 176–81, 182, 184–85, 187; IQ and, 169; in research design/interpretation, 238; self-identification, 180–81, 185, 188–89

Rajagopalan, Ramya, 50

ramets, 60–63

Rapoport, Anatol, 109

RF-ACE (Random Forest-Artifical Contrasts with Ensembles), 75–77, 80–81, 82

Rheinberger, Hans-Jörg, 37

Risch, Neil, 157

RNA transcripts, 18–21, 22, 66, 237; role in evolution, 23–24; understanding of, 123

Roe, Bruce, 47

Rose, Nikolas, 91

Rose, Richard, 166

Rotimi, Charles, 184

Ruaño, Gualberto, 179

Rutter, Michael, 160

Santa Fe Institute, 107

self-organizing maps (SOM), 96–98, 97

Shapin, Steven, 186

Shenassa, Edmond D., 204

single-nucleotide polymorphisms (SNPs), 1, 21–22; associated traits, 83, 85–87, 87, 88, 90–91, 151, 159; dimensionality, 93; lasso technique, 89, 90, 100; single point mutations, 23

smoking, research on, 166

SNPs. *See* single-nucleotide polymorphisms

SOM. *See* self-organizing maps

Somel, M., 23

South African Human Genome Programme, 187

Stengers, Isabelle, 100

Stotz, Karola, 214–15

Strasser, Bruno, 82–83

Strogatz, Steven, 109, 110, 111

Surgeon General's Office, 177

Synthetic Cohort for the Analysis of Longitudinal Effects of Gene-Environment Interactions, 183

systems biology, 83, 99, 110–11, 121–22, 214, 215, 233, 235

Szyf, Moshe, 213–14

Taft, Ryan J., 22

Tatum, Edward, 14